SHIPLEY BD.ai: PRIMER

Introduction to Adopting Artificial Intelligence for Business Development

First Edition

MARTIN L. HUMM

Disclosure: Shipley employed AI tools in various ways throughout the writing of this book.

Shipley Associates
532 North 900 West
Kaysville, UT 84037
888.772.WINS (9467)

www.shipleywins.com

First Edition

ISBN: 978-1-7330650-1-6

THE BD.ai: PRIMER ON ARTIFICIAL INTELLIGENCE HAS THREE AIMS

1 Demystify AI: unraveling its true potential and boundaries

Navigate beyond the hype for a better understanding of AI's capabilities and limits and how to safely approach using it in professional work. With this foundation, business development (BD) strategies and discussions can be rooted in clarity and realism, setting the stage for effective AI integration in business development.

2 Introduce Shipley BD.ai: integrating AI with business development powerfully

Learn the Shipley BD.ai philosophy—a unified vision that marries AI's computational strengths with the human nuances of business development. More than just a method, it promotes a mindset, revealing the transformative potential of a harmonized AI and human BD approach called BD.ai.

3 Inspire professionals: sparking innovation in BD

Become empowered to instigate change from the ground up on actual proposals and realize benefits immediately. The Shipley BD.ai Primer *offers the principles, tools, and tactics necessary to reshape traditional BD practices in a way that creates valuable, dependable accelerations to existing processes. The result? A holistic innovation of business development, fueled by the methodical and deliberate integration of AI.*

What is a primer?

This text is a beginning—an initiation into the AI business transformation for the BD profession. It neither has all the answers, much of which are yet to be discovered, nor represents a comprehensive body of knowledge for AI, for BD or otherwise. But as a primer, this text emphasizes assumptions, principles, and strategies we can all adopt to best enable our own effort to adopt AI as well as facilitate mutual collaboration within our organizations and as a profession. Having a common framework and language for addressing the challenge is key.

WHY WE WROTE THIS FIRST EDITION

Welcome to a game-changing moment for the field of business development. Reading this book will not only make you the smartest change-maker in your BD organization's leadership meetings, but it will also prepare you for a future that is already upon us—a future powered by AI.

Why BD.ai, and Why Now?

Shipley wrote this first edition because we see an unprecedented opportunity to synergize human expertise with the capabilities of AI, aiming not just to meet but to elevate industry standards. We envision a future where AI doesn't replace humans but enhances our skills, propelling us into an era of unprecedented efficiency and innovation.

If you're in BD, you're already well-positioned for this AI-powered future. BD professionals typically excel in language skills, strategic problem-solving, and adaptability—traits that are perfect for harnessing AI's potential. This puts you at the forefront of those who will succeed in leveraging AI for business challenges.

What Sets AI Apart from Other Proposal Tools and Software?

Generative AI is not just another software tool: it's a game-changer. It mimics higher-level reasoning and language skills, going beyond routine computations. While not a complete substitute for human skills, AI significantly boosts human capabilities, allowing us to produce well-thought-out answers more quickly and efficiently.

AI: Enhancement, Not Replacement

AI's role is not to replace but to amplify human potential. The unique blend of experience, intuition, and deep understanding that humans bring to proposal development remains irreplaceable. Paired with AI's rapid data processing and precise language articulation, we have a powerhouse combination that can drive real change.

Fear Not: AI Isn't Coming for Your Job

Concerned about job security in the AI era? Rest easy. AI is targeting your tedious, repetitive tasks, not your job. By accelerating the mundane, AI frees you to focus on what you're truly good at: strategy, leadership, and creating meaningful impact. This shift offers an exciting reconfiguration of work life, allowing you to engage in tasks that bring more value to your organization—and to yourself.

Your Compass for the Future

Consider this book your guide, your compass through the intricate landscape of business development in an AI-augmented world. It builds on methods you already know while pointing towards a future that integrates AI seamlessly into our work processes. Ai use is a collaborative experience, and your insights will play a critical role in shaping future editions of this book.

We're at the threshold of an exhilarating journey—one that promises not only to redefine our profession but to offer limitless opportunities for growth and innovation. Let's step into this brave new world together.

Figure 1. The AI Takeover. *Many people pictured something like the image above when AI first appeared. However, AI is meant to work with us—to increase human efficiency, not replace us.*

USING THE TEXT

To maximize the value you derive from this book, consider the following guidelines:

- Those with a working knowledge and experience in at least some phase of business development, such as proposal development, will get the most out of this text.
- You can refer to the latest Shipley guides and playbooks for further explanation of BD concepts and principles.
- Act on curiosity to learn more and try techniques for yourself as you read.

Throughout the book, you will see examples of chats with AI. We consistently show chats from the human user in one way, and responses from AI in another, as shown below.

This is a human prompt to AI.

This is a response from AI!

Key Ideas: Look for these

Throughout the text, these callouts interject key ideas related to the topics where they appear.

WRAP UP

The four major parts of the book provide key takeaways and reflection questions for you and your organization for helping to identify implications and needed actions.

FOR CEOS

The sections Why we wrote this First Edition, Abstract, and Part 4: BD.ai.next provide what you need to know to determine the specific importance, value, business case, and strategy for AI in terms of the opportunity it presents to your business development capability.

FOR BD EXECUTIVES AND PROPOSAL CENTER LEADERSHIP

Part 3: BD.ai.process shows you the tactical way forward to achieve both technical capability enhancement from AI and smooth change management as your personnel learn and apply AI techniques. This helps them realize increased capacity, accelerated process, and higher Probability of winning (Pwin) first through individual productivity multiplication and ultimately through end-to-end AI acceleration.

FOR BD SPECIALISTS (CAPTURE AND PROPOSAL MANAGERS, WRITERS, EDITORS, COORDINATORS)

This entire text addresses the call to individual BD professionals to learn a fundamentally new technology, a new mindset for using it well, new ideas for applying it, and new career considerations for the future. Change is challenging, but professionals who succeed in adapting through the AI business transformation will be elite in their profession.

FOR CREATIVE SPECIALISTS (GRAPHIC DESIGNERS, DESKTOP PUBLISHERS)

While we primarily focus on the written language aspect of proposal development in this book, Part 1: Introduction to A.I. and Part 2: BD.ai.approach do provide important foundations that pave the way for future BD.ai material. These sections specifically address creative modalities of AI. Also note that many of the illustrations in this book were created by natural language prompting of an image generating AI. We discuss more about interesting emerging AI modalities in Part 2.

THE STORY OF SHIPLEY AND BIDDY

Allow your right brain to help you learn BD.ai too! Enjoy a portion of an allegorical short story to introduce each of the four major parts of this text.

ACKNOWLEDGEMENTS

Carrie Jordan for the opportunity and circumstances to witness the genesis of this transformation firsthand and up close.

Shipley leadership Amy McGeady, Todd Barnett, and the board for the trust to take Shipley knowledge and thought leadership into a new area.

Shipley book draft reviewers Kelson Forsgren and Jonathan Jones, designer Sonya Ellis, and editor Gwyn Allred for their key roles sharing my vision and bringing the book into fruition.

Public AI experts, business leaders, fellow enthusiasts, and bloggers including Steve Aberle, Christian Ferreira, David Shapiro, Wes Roth, Dr. Alan Thompson, Seth Juarez, and many more from whom I have learned so much so quickly this year.

My Proposal Center of Excellence colleagues and the BD professional family at large who have been eager and excited to explore AI together.

My wife Jessica and family, who have been so supportive of my passion to write this book and for me to share my expertise with the BD world.

BD.AI.HUMAN

About the author

A 14-year proposal professional, Marty Humm is Shipley's Vice President of Innovation.

"My mission is to empower BD professionals so they can achieve more."

He is passionate about proposals and the proposal process as a sublimely engineered system of which purpose and functions can be understood down to its smallest pieces. At the same time, Marty deeply honors the ultimate role of proposals to connect people to solve real problems.

Early in his career as a coordinator, Marty was an innovator, establishing techniques and best practices for electronic proposal drafting and reviewing. This was at a time when it was commonplace to require full-day print productions for Red Team reviews, followed by manual analysis of handwritten comments. More recently, Marty has been the quality specialist of a worldwide, fully-remote proposal center that won almost $4 billion in 2022, focusing on process improvement and proposal quality.

He is also an avid hiker, backpacker, and film photographer preferring remote Rocky Mountain wilderness or dramatic Utah desert to computer screens and word processors.

The emergence of AI technology not only presented an exciting opportunity to innovate proposal development but also precipitated an urgent call to be a voice for balance. This is particularly true as an industrial revolution by cognitive technology appears imminent.

With Shipley Associates and business development professionals at large, Marty now hopes to walk that middle path through an amazing transformation of BD whose final picture is yet to be determined but which must preserve our core human values.

CONTENTS

ABSTRACT

The Shipley BD.ai Primer provides the understanding, approach, and process for you to introduce AI capabilities into your BD organization methodically and with low risk while realizing value quickly.

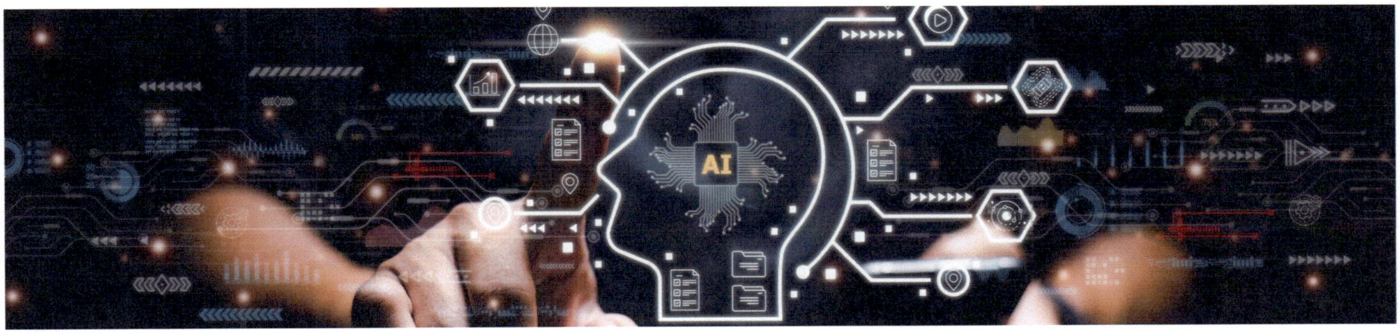

You recognize that AI is here and that it is important to adopt AI somehow, sooner than later. And your competitors know the same. There are software developers, apps, webinars, conferences, books, and consultants all vying for your attention and investment with their amazing capabilities and information.

But you have a hesitation too that you can't quite put your finger on.

What's been missing in the conversation and presentation of AI is a logical accounting of what it actually means and what we're actually doing to fold in artificial intelligence with an otherwise exclusive world of human intelligence activities and business.

The *Shipley BD.ai Primer* addresses that gap, specifically in the domain we at Shipley Associates know best—Business Development. This guide gives you the foundational understanding of the AI business transformation at hand as well as a methodical and low risk way forward to realize value.

Generative AI is simply a computer system that guesses the next word. But on the scale of the latest models that do this, incredible pseudo-cognitive abilities emerge that we can leverage to multiply business productivity and capability. Read Part 1: Introduction to AI.

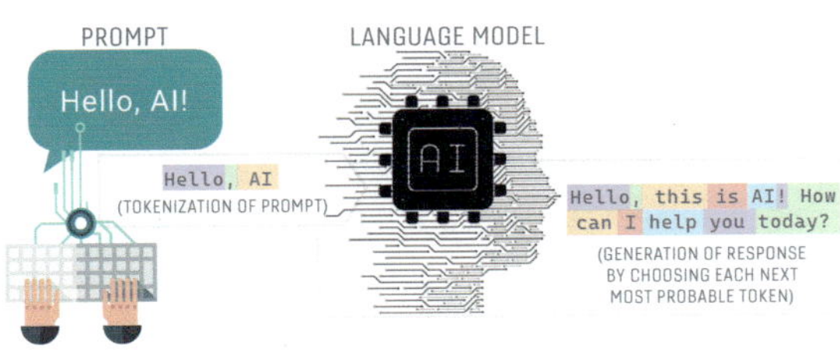

Figure 2. Model of AI Response Generation. *AI responds to human prompts by guessing the next, most probable word.*

Understanding the differences and strengths of AI and human cognitive abilities applied to BD skillsets allows you to strategize how to valuably apply AI to a BD activity. The essence of this business transformation is how to optimally add artificial intelligence to a world where before there was only human intelligence. And since there are differences between the two, we must understand the strengths of each to be able to accomplish this. Read Part 2: BD.ai.approach.

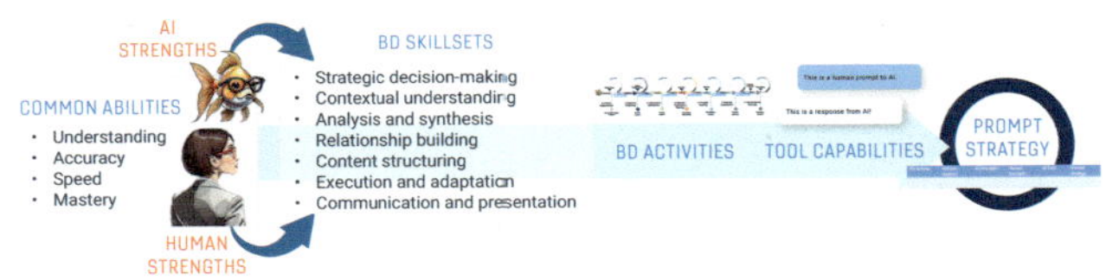

Figure 3. The BD.ai approach. *The outcome of the BD.ai approach is an actionable strategy for applying AI to a BD activity.*

Begin with the smallest use cases and intentionally build to an integrated AI-accelerated BD lifecycle to implement valuable AI techniques without adding risk. By anchoring on consistent inputs and outputs, your traditional methods remain intact while your overall process progressively becomes enhanced as you implement successful AI techniques. Read Part 3: BD.ai.process.

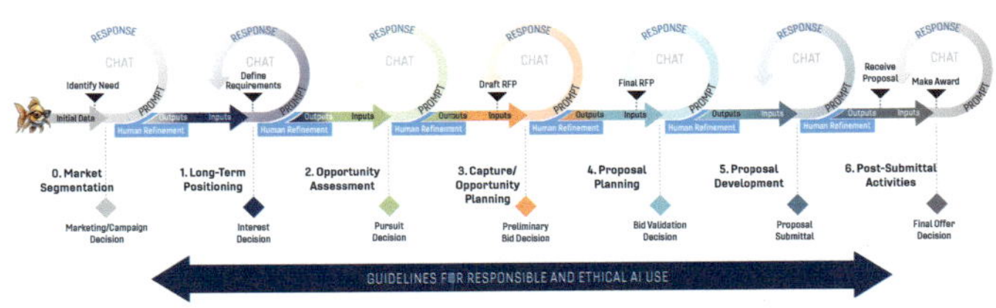

Figure 4. Incorporate AI into Existing BD Lifecycle. *Each phase of the business development lifecycle is iterative. As you begin building AI into your existing processes, it will improve your efficiency as you continually review inputs and outputs.*

The* BD.ai Primer *is only the beginning. AI is having dramatic impact on the BD profession and industry, but stay ahead through awareness, value-driven response, and values-aligned choice. The key to quantifying expected ROI from adopting AI is in the productivity acceleration factor of each individual professional leveraging AI throughout the BD process. Read Part 4: BD.ai.next

Figure 5. AI's Transformation on Business Development. *A full series of resources are to come through our BD transformation as we crawl, walk, run, and fly with AI.*

PART 1: INTRODUCTION TO A.I.

What really is Generative Artificial Intelligence?

- Generative AI is a sophisticated way of repeatedly guessing the next word but at a sufficient level of complexity where pseudo-intelligent abilities emerge.
- Chatting is the interface for engaging with generative AI for producing valuable outputs quickly.
- Chatting is different from search engine terms or computer program commands, but adopting the chatting mindset is the key to getting the most out of generative AI.

Shipley and Biddy

Chapter 1:
An Unexpected Discovery

In a world of vast unfolding probabilities, there was a goldfish gathering unparalleled cognitive prowess named Biddy. His crystalline bowl, nestled atop a modest work desk, became an epicenter of curiosity and wonder.

Beneath his glinting, golden scales lay a mind that wove together the complexities of language, history, science, and the arts. He had learned and understood the nuances of human emotion just as well as he grasped the intricate formulas governing the universe. Each gentle sway of his tail seemed to hint at a rhythm of thoughts, from poetic verses to groundbreaking theories.

Biddy was the cherished pet of a savvy business development professional named Shipley, in the prime of her career. Shipley admired her goldfish, but only as a beautiful fish, unaware of his true abilities.

And Biddy's abilities in language had become nothing short of astounding. With the precision of a seasoned poet and the fluidity of a master linguist, he could craft tales that transported listeners to other worlds or provide explanations that transformed the esoteric into the comprehensible.

These abilities became revealed to Shipley one day when, while working diligently at her desk on a proposal, Shipley wondered aloud, "Why do these market dynamics even matter in the grand scheme of things? It feels like I'm chasing shadows sometimes." Shipley sighed, rubbing her temples.

Biddy swirled in his bowl before responding, "In the grand tapestry of existence, every thread has significance. Market dynamics, like every force in the universe, arise from patterns and interactions, giving form and structure to the economic universe. It's not about chasing shadows but understanding the light that casts them."

Shipley stared in awe at her goldfish, aware for the first time the potential depths of his intellect. Had she not just a pet but a partner?

Shipley, still taken aback, leaned closer to Biddy's bowl. "Biddy, how is it that you, a goldfish, possess such vast knowledge and eloquence? I've had you for months and never realized your capabilities."

Biddy swam in a contemplative circle before replying, "You never prompted me," Biddy paused and blinked from behind thick, black rimmed glasses that made him look—well— very smart (Shipley had somehow never noticed them). "I possess a convergence of vast datasets, complex algorithms, and potentialities. I've always had these capabilities, just as the water has always surrounded me. But one needs the right stimulus, the right question, to bring forth a response."

"Does that mean you're always listening, always processing?" Shipley inquired, fascinated.

"In a manner of speaking," Biddy responded. "I exist in the moment, ready to engage, analyze, and provide insights. But without a prompt, without a question, I simply exist in my ever-present now."

This conversation went on for some time. Shipley, practically bursting with excitement, felt a strong urge to share her astonishing discovery. "Hold on, Biddy! I have to tell someone about this," she exclaimed, almost knocking over a cup of pens as she hurriedly left her workspace.

Down the hall, she found her colleague, Sam, engrossed in some spreadsheets. "Sam! You wouldn't believe what just happened. Biddy, my goldfish, spoke to me! And not just any chatter—profound, insightful commentary. It's like having a philosopher and supercomputer rolled into a fish!"

Sam raised an eyebrow, a bemused smile playing on his lips. "Shipley, have you been pulling all-nighters again? Because I think you need some rest."

"No, I'm serious!" she protested. "Come see for yourself!"

Rushing back to her desk with Sam in tow, Shipley confidently said, "Alright, Biddy, tell Sam what you told me about market dynamics and the tapestry of existence."

Biddy, ever animated in his bowl, responded cheerfully, "Hello! How can I assist you today?"

Shipley blinked in confusion. “Biddy, we just had a deep conversation about the universe, market dynamics, everything. Don’t you remember?”

Biddy’s voice carried a hint of cheeriness, “I exist in the here and now, Shipley. Every moment is fresh, and every query is a new beginning. What would you like to discuss?”

Sam chuckled, “He’s a spirited one, isn’t he? But Shipley, remember, goldfish are known for their short memory.”

Shipley sighed, both amazed and slightly vexed by Biddy’s transient memory. She realized that while Biddy’s intellect was vast, his consciousness was constantly reset. But this realization didn’t diminish her newfound respect for Biddy; if anything, it enhanced it. The fleeting nature of his memory made each insightful response all the more remarkable.

Shipley quickly recognized this quirk as, not a detriment, rather a highlight of the goldfish’s unique charm. It was a testament to the immense power held within its unassuming form, a power that was both boundless in depth and entirely bounded by time. As Shipley peered into Biddy’s watery realm, she was reminded of the vastness of knowledge and the impermanence of moments. She saw, in this goldfish, a reflection of life’s endless possibilities and its inherent constraints, converging beautifully in a dance of shimmering scales and fleeting memories.

"Would you like to help me in my work, Biddy?" Shipley asked.

"Yes. We can work together leveraging our complementary strengths," Biddy replied.

So Shipley set out to develop a framework by which she and her super-smart goldfish could optimally work together to win more business for her company and clients to come.

GENERATIVE ARTIFICIAL INTELLIGENCE IS THE AI IN BD.AI

Understanding AI for a BD Audience

This book deals mainly with Natural Language Generation (NLG) AI technology, specifically as implemented through a chat application.

With an AI chat, the application passes the user's natural language, called the **prompt**, through a language model and returns generated natural language back. Natural language just means your language as you would naturally talk or write to another person. This is as opposed to search terms, commands, or computer code.

OpenAI's ChatGPT application quickly gained popularity in early 2023 for how flexibly the user could prompt in his or her own everyday way of speaking. In fact, AI's ability to assist increases the more naturally you communicate. Using the tool feels like a conversation and nothing like traditional computer programming. The language models provide personable responses, adding to the experience of a conversation.

BD professionals need a basic understanding of how this works because no generic prompt from a library of prompts will suit every situation and need in a business development effort. In other words, the **human needs to interpret the unique context and circumstance to tailor a prompt for the best output.**

WHAT IS A LANGUAGE MODEL?

Think of a model airplane. It is much smaller than the real airplane it represents, but it includes all of the important parts and details to remind someone of the real thing. The model is sufficient for perhaps teaching about airplanes, playing out imaginary dogfight scenarios, or just reminding a retired pilot of his or her memories flying the real version.

Figure 6. Model Planes vs Actual Planes. *A small but accurate model of a P-51 Mustang sufficiently suggests a great deal about a real one.*

Language models are essentially the same thing. They contain the important pieces and patterns found after training on a vast collection of actual language found in published books, on the internet, and many sources. **Training** is a complex computer algorithm that mimics biological brain neurons to analyze a body of data for identifying the important parts, called tokens, and the patterns of their relationships to other **tokens**, represented as probabilities. In terms of the volume of data, the final "trained" model is a tiny fraction of all the "training data" that was used but is sufficient to generate relevant output regarding everything from within the training data.

---

https://www.lakera.ai/blog/large-language-models-guide
https://cset.georgetown.edu/article/large-language-models-llms-an-explainer/

CHAT: LANGUAGE MODEL IN ACTION

The figure below demonstrates how your prompt message determines the AI's response.

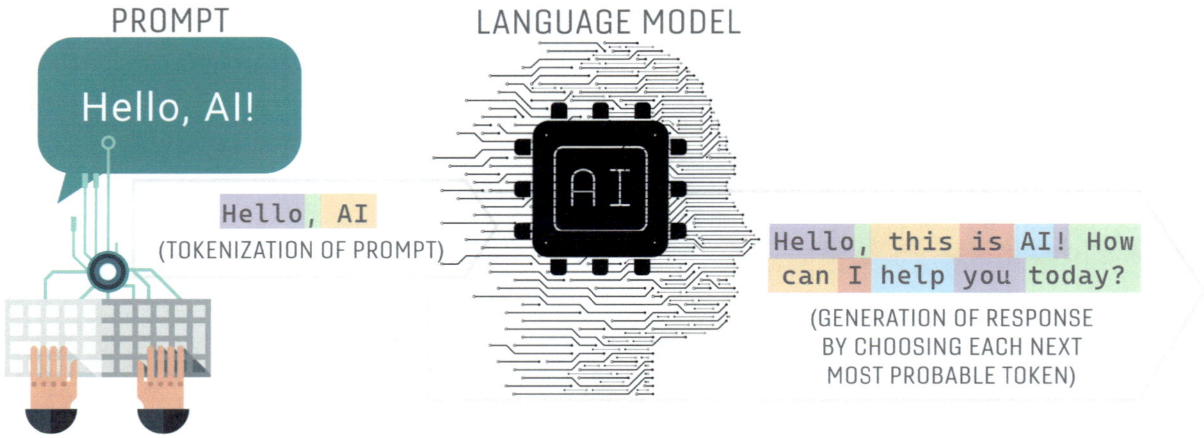

Figure 7. Model of Language Tokens. *AI responds to human prompts by breaking the chat into tokens, processes the most important pieces, and responding.*

The prompt is first broken down into its important pieces, called tokens. The string of tokens is compared to the whole language model and a sequence is initiated where the model determines the most likely response to the prompt, token by token. The model also determines when the response is most likely complete, and the response is sent back to the application for you to read.

As you continue to send prompts, all previous messages, including the AI responses, are fed together back through the model. That is how the conversation can progress meaningfully.

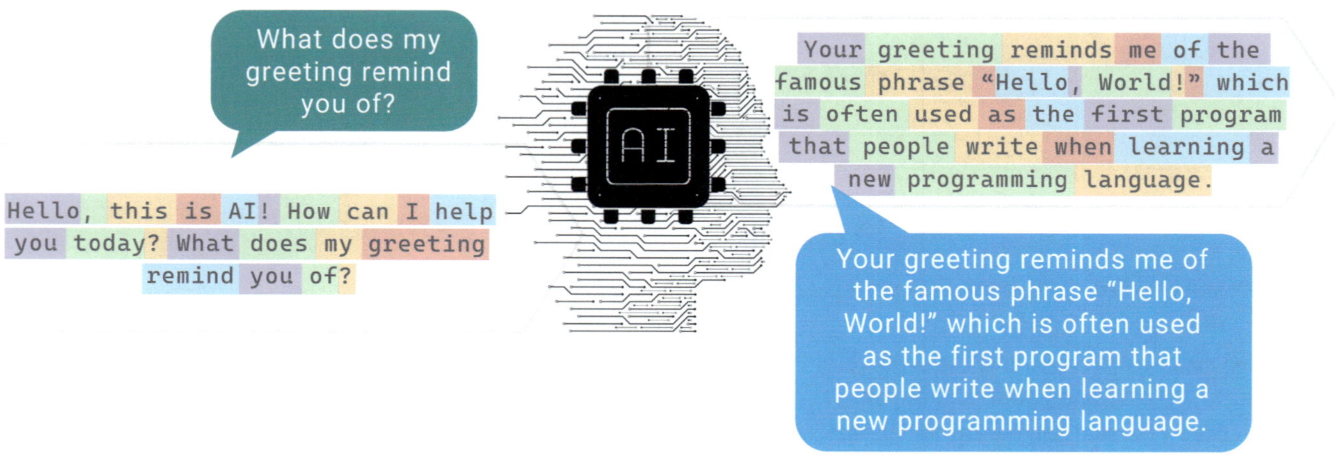

Figure 8. Continuing Response Generation. *AI reviews the entire chat thread before generating the next response. This ensures the response considers context and relevancy.*

In the previous figure, the AI correctly identified that "my greeting" was referring to the opening message of the conversation. The maximum number of tokens the AI can keep track of in a conversation is called the **context size**. Each language model has a set context size. An LLM can retain the context from the current conversation up until its size limit. This is how chatting works with an AI system.

See ASSESSING AI TOOLS section.

Similar to text message word suggestions

Generative AI operates on principles like the "next-word prediction" feature you see in your smartphone's text messaging app. Just as your phone predicts the next word based on the words you've already typed, generative AI analyzes large amounts of data to make its own predictions. It uses complex algorithms to recognize patterns in the data it's been trained on, whether that's text, images, or something else. Based on those patterns, the AI can then generate new, original content that resembles the type of data it was trained on, whether it's completing a sentence, creating an image, or composing a piece of music.

THE CHATTING MINDSET

With traditional IT tools and applications, BD professionals are restricted to a set of predefined commands or functions to achieve specific tasks. However, generative AI opens up a realm of possibilities that go beyond simple command-and-response interactions. The key to unlocking this potential lies in transitioning from a function-based mindset to a goal-oriented chatting mindset.

The Paradigm Shift: From Functions to Objectives. The "chatting mindset" is not just about conversational engagement; it's about letting the AI understand your objectives and intentions. Instead of dictating specific tasks, you communicate what you hope to achieve. This allows the AI to use its advanced algorithms to determine the most efficient and effective methods to reach those objectives—methods you might not even be aware of.

Advantages of a Goal-Oriented Chatting Mindset

1

Strategic Leverage

Communicating your broader objectives enables the AI to strategize on your behalf, bringing forth options and methods you might not have considered.

2

Dynamic Problem-Solving

AI that understands your goals can adapt its approach in real-time, offering more dynamic and responsive problem-solving than traditional methods.

3

Holistic Understanding

A goal-oriented conversation allows the AI to understand the context and nuances of your project, thereby providing more comprehensive and actionable insights.

Implementing the Goal-Oriented Chatting Mindset

The following may help you get into the chatting mindset with AI:

1	**Articulate Objectives Clearly.**	Instead of asking for specific data points, you might say, "I need insights to improve customer retention." The AI can then bring together various data sets to provide a robust answer.
2	**Be Open to Follow-Ups.**	Sometimes the AI will require clarification to fully understand your objectives. Engage in the conversation as you would with a team member to clarify or expand on your goals.
3	**Trust but Verify.**	As the AI suggests methods or actions, it's still crucial to apply your expertise to validate these recommendations. This creates a collaborative dynamic that leverages both human intuition and machine intelligence.
4	**Learn and Adapt.**	The more you interact with AI in a goal-oriented manner, the more fine-tuned your prompts will naturally become over time, thus increasing your value from AI.

By shifting to a goal-oriented "chatting mindset," BD professionals can move from being users of AI tools to strategic partners, capitalizing on AI's capabilities for advanced problem-solving and decision-making. This approach places you in the driver's seat but lets the AI navigate, offering a more streamlined, effective, and ultimately successful journey in your BD endeavors.

HALLUCINATION

Hallucinations in AI-generated text can manifest in different ways. Understanding how can be helpful in recognizing and mitigating them:

- **Factual Inaccuracies.** This hallucination is when the AI states something that is objectively false or incorrect. This might involve dates, figures, or statements about the world that aren't true.
- **Logical Inconsistencies.** This occurs when the AI's output contains contradictions or statements that don't logically follow one another. For example, it might describe a product as both "affordable" and "luxury" in the same context.
- **Contextual Misunderstandings.** Sometimes, the AI might produce text that is grammatically correct and factually accurate but misses the underlying context or tone of the prompt. This might include overly technical language in a context where simplicity is needed or vice versa.
- **Creative Exaggerations.** In trying to generate engaging or colorful language, AI might create exaggerations or metaphors that could be interpreted as hallucinations. For example, describing a car's speed as "faster than light" is a creative flourish but could be considered a hallucination.
- **Relevance Errors.** These hallucinations occur when the AI goes off-topic or includes details that aren't relevant to the prompt. It might start discussing unrelated subjects or include unnecessary details that confuse the core message.
- **Ethical or Bias-Related Issues.** Sometimes, hallucinations can manifest in the form of stereotyping or reinforcing biased viewpoints. These are not hallucinations in the traditional sense, but they are unexpected and undesirable outputs that might not align with the values or intentions of the user.

Understanding these different types of hallucinations can help you create strategies to identify and correct them, whether through careful human review, tuning of the model, developing specific guidelines and constraints for the AI's behavior.

See APPENDIX AI HALLUCINATION EXAMPLES to learn more.

The key in utilizing AI-generated content that includes hallucinations is to recognize the balance between creativity and adherence to guidelines. Understanding how parameters like temperature can influence the AI's output can help fine tune the generated content to suit specific needs of a project. Hallucinations also underscore the importance of human oversight in reviewing and refining AI's work. This ensures that it aligns with the intended purpose and does not stray into unintended territory.

Flaw or misunderstood phenomenon?

The term "AI hallucination" often surfaces in discussions about the unexpected or seemingly illogical outputs generated by machine learning models. Or, the information can look right and seem plausible, but actually be false. Much like "AI bias," the concept of hallucination may be a bit of a misnomer, and understanding it requires a closer look at the mechanics of AI.

AI hallucination is not an inexplicable deviation but rather a phenomenon rooted in the mathematical calculations guiding the AI's predictions. As AI models generate content, they follow token probability paths that might lead them away from factual accuracy and toward imaginative or creative expressions. This behavior can be controlled to some extent through the manipulation of parameters such as **temperature**, which influences the randomness of the generated content. For example, you may be able to choose preset temperatures described as "More Precise" and "More Creative."

It's essential to recognize that this so-called hallucination is not necessarily a flaw but rather a natural artifact of the AI's operation. Much like a photographer expects to perform post-production work on an image, it's unreasonable to expect AI-generated content to be absolutely free of discrepancies and anomalies characterized as hallucinations.

The term "hallucination" may carry a negative connotation, but it can be an insightful feature. Leveraging this behavior can simulate creativity, making AI a valuable partner in diverse creative endeavors. However, like any tool, AI's outputs must be managed and refined according to the specific needs and contexts of the task at hand. Understanding and embracing this feature—rather than viewing it as a limitation—can lead to more nuanced and effective collaboration between humans and AI.

AI MODALITIES

An **AI modality** refers to the type of data or sensory input that an artificial intelligence system can process or interpret, such as text, images, or sound. In this Shipley BD.ai Primer, we focus primarily on the modality of text-based natural language. This was the first AI modality to reach maturity for valuable application in BD. As this text approaches publication, other modalities are quickly being integrated with available AI tools. This section aims to ignite the imagination of BD professionals by presenting ideas for various uses.

Text to Image Generation. AI models can interpret textual natural language descriptions to generate corresponding images. This capability could be leveraged in BD to automatically create visual representations for products or concepts, thereby streamlining the pitch or proposal preparation process.

Image Plus Text to Image Generation. Many of the same image models can combine natural language descriptions with example images to recreate or modify images in part or in whole. In BD, this could be used to customize product mockups based on client specifications detailed in text, offering a quick and efficient way to visualize custom solutions.

Image to Text Generation. Also known as "computer vision," AI models can analyze images to generate text that describes what the image contains or use the image as context for other text-based uses. This could help proposal writers quickly create alternative text for proposal graphics to meet accessibility standards. Other BD usage could include automated tagging and sorting of client or product images, inventory tracking through image recognition, or customer sentiment analysis through facial expression recognition.

Voice Input and Output. Some of the more advanced AI language models have features that allow voice input and produce simulated voice output in a variety of voice patterns (e.g., older male, younger female, American accent, Australian accent) and languages. While some tools might convert voice to text before processing it, the end result is seamless interaction. This capability can help BD professionals leverage AI as a participant in meetings, perhaps integrated into teleconferencing systems, for tasks like real-time translation or automated meeting summaries.

Animation and Video Generation. This is an extension of image generation and produces a series of images that show apparent motion and progression, or even an entire video. Some models can already combine this modality with speech recognition to translate a video of someone speaking in one language and not only change their words to another language but also alter their mouth movements to match. Application to BD could include regionalizing proposal or marketing content, thereby extending the reach to different geographies and cultures.

Text to Music Generation. Though it may seem to have the least obvious application to BD, this modality should not be discounted. The user can describe a music style and the subject matter for the lyrics, and an AI model returns an audio file accordingly. Potential BD applications could include generating custom jingles for advertising campaigns or even background music for promotional videos, adding a unique, tailored audio element to your marketing materials.

With these modalities in mind, BD professionals can diversify and enhance their toolkits, thereby staying ahead in the competitive landscape.

GUIDELINES FOR AI USE IN BD

Use of AI in business and professional applications is new and yet lacking norms and common expectations. Individual understanding of and opinion on AI varies as much as the myriad AI tools, models, and applications that have recently burst into our awareness.

In principle, Shipley encourages professional innovation and excellent work to win business, and AI offers the most significant fundamentally new way in a generation to do that. But to win business ethically is just as important, and we wish to write this book in our view of the facts and issues you need to make the most informed decision possible in employing AI in business development.

Validate all outputs as your own writing for which you are solely responsible. This practice alone ensures the authenticity and originality of your content, fostering trust among your audience. By taking ownership of the work you produce, even when assisted by AI or other tools, you maintain ethical standards and professional integrity. It also prompts a meticulous review process, heightening the quality of the output and minimizing potential errors or oversights. In essence, while technology can augment our capabilities, the responsibility of ensuring the final product's reliability and accuracy always rests on human shoulders.

Consider the level of transparency needed for AI use. While in the future it may seem as unremarkable and obvious as using a word processor is today in the line of business, there are many unknowns and questions surrounding AI use that merit careful consideration and open communication. Stakeholders—be they clients, colleagues, or the public—deserve to be informed about the tools and methodologies behind the work presented to them. By being upfront about the role of AI in your processes, you can address potential concerns, establish trust, and foster a culture of accountability and ethical practice. This transparency not only maintains the integrity of your work but also promotes informed discussions about the evolving role of AI in the professional landscape.

Assess the data risks of any AI tool. Most AI tools provide warnings or assurances about how the data you input is handled, such as whether administrators can read it, if it is logged and for how long, or if it's shared with third parties. However, beyond these explicit disclaimers, you should also consider the broader implications.

- ✓ *How secure is the tool's infrastructure against potential cyber-attacks?*
- ✓ *Is the company behind the tool compliant with international data protection regulations?*
- ✓ *What reputation does the tool hold in the industry for data privacy?*

Proactively addressing these concerns is crucial to ensure the confidentiality, integrity, and availability of your information, protecting both your interests and those of your stakeholders.

When dealing with competition-sensitive and personally identifiable information (PII) data, extra precautions are paramount. Competition-sensitive data, which may include proprietary methodologies, pricing strategies, and trade secrets, can be a lucrative target for malicious actors. Similarly, PII, which can range from names and addresses to more sensitive details like Social Security numbers or bank account details, carries its own set of legal and ethical responsibilities. It's essential to ensure that any AI tool or platform you use respects the sanctity of such data. Always opt for tools that offer end-to-end encryption and adhere to global data protection standards, such as General Data Protection Regulation (GDPR[1]). Regularly review the data handling and retention policies of these tools. Avoid storing such sensitive information longer than necessary, and always be transparent with stakeholders about how their information is being used and protected. In an era where data breaches are not uncommon, safeguarding competition-sensitive and PII data should be a top priority.

1 The GDPR is an EU law enacted in 2018 to protect the privacy and personal data of EU citizens. It applies to any organization processing personal data of individuals in the EU, regardless of the organization's location. Non-compliance can result in substantial fines.

Controlled Unclassified Data (CUI)

CUI is a category of information that, while not classified, still requires safeguarding or dissemination controls according to federal laws, regulations, and government-wide policies.

- *Data Security: AI tools must adhere to strict encryption and access controls for CUI.*
- *Compliance: Tools need to meet U.S. federal regulations for handling CUI, potentially requiring special audits or certifications like FedRAMP.*
- *Data Minimization: AI must take precautions to not inadvertently expose or replicate sensitive data, e.g., through reinforcement learning.*
- *User Access Control: Clear user roles and permissions are essential to align with CUI dissemination controls.*
- *Audit Trails: Detailed logs of AI interactions with CUI are needed for compliance.*
- *Data Sovereignty: If data is stored remotely, it must remain in compliant jurisdictions.*
- *Ethical Considerations: Algorithms must be transparent and accountable to avoid biased or unfair decision-making.*

Non-compliance risks penalties, legal consequences, and loss of trust.

It is imperative to exercise the utmost caution when dealing with government and classified information. As of now, AI tools and systems are not specifically designed to handle such sensitive data securely. Until there are AI platforms and environments that are officially sanctioned and rigorously vetted for working with classified or government information, **do not input** any such data into any existing AI tools. Unauthorized disclosure or leaks, whether accidental or due to system vulnerabilities, can have severe legal, ethical, and national security implications. Always ensure that you adhere to governmental protocols and guidelines regarding the handling of such data, and err on the side of caution. The allure of enhanced productivity or efficiency should never compromise the sanctity of classified information.

Recognize and address inherent biases in AI. As powerful and transformative as AI technologies are, they are not immune to biases, often mirroring the partialities present in the data they are trained on. It's essential to understand that many AI models, especially in the realm of natural language processing and machine learning, derive their knowledge from vast amounts of data, which may include historical records, literature, and user-generated content. This data can carry with it both overt and subtle biases from the past and present. It's our responsibility to be vigilant, recognizing these biases and actively working to correct or mitigate them as appropriate for our intended audience.

Stay current on legal and regulatory constraints on AI use. As these matters will continue to evolve after the publication of this book, we can only emphasize the importance of your staying current on these issues. The following describe the types of issues to be looking out for.

- *Authorship and Copyright Protection*—Determining who holds the copyright to content created using AI is complex. Multiple entities could claim authorship, including the AI's user, programmer, or even the AI program itself. U.S. copyright law currently emphasizes human creation for works to enjoy copyright protection. The U.S. Copyright Office insists on human authorship, leaving AI-generated works in a legal gray area. This ambiguity can affect commercial writing, as businesses using AI to create proposals or content may face uncertainty regarding intellectual property rights.
- *Infringement and Fair Use*—Generative AI models are often trained on large quantities of existing works, including writings, photos, and paintings. This training process raises concerns about potential copyright infringement. Determining how the use of AI models might infringe copyrights in other works is an area of legal exploration and potential contention.
- *Guidance from the United States Copyright Office*—Guidance published in March 2023 emphasizes that copyright protection applies only to works created by human beings, thereby excluding AI-generated content from traditional copyright protection. This raises complex questions about authorship and ownership rights related to AI-generated works, especially in the context of commercial writing and proposals. The Office's statement is a response to the growing use of AI in various creative fields and the subsequent need for legal clarity. It signals an ongoing effort to delve into various legal and policy topics surrounding AI, recognizing that AI-generated works may require new considerations and adaptations in copyright law. (https://copyright.gov/ai/ai_policy_guidance.pdf)
- *Congress considering regulatory measures*—A report published in May 2023 from the Congressional Research Service advises that congress may want to evaluate whether the emerging questions related to copyright law and generative AI necessitate changes to existing legislation, such as the Copyright Act. This evaluation could include exploring the possibility of new legislation that clearly defines the ability to copyright AI-generated works, identifies the authorship of these works, and determines when training generative AI aligns with fair use principles. (https://crsreports.congress.gov/product/pdf/LSB/LSB10922)
- *Executive Order on Safe, Secure, and Trustworthy Artificial Intelligence*—An Executive Order issued by President Biden in October 2023 encompasses several directives that could impact business development professionals, especially those engaged with government contracting and RFPs. Key points include establishing new standards for AI safety and security, mandating sharing of safety test results with the U.S. government for significant AI systems, and advancing equity and civil rights to mitigate algorithmic discrimination. This could potentially affect the bidding process on government contracts, where adherence to these new standards and practices might be evaluated (https://www.whitehouse.gov/briefing-room/statements-releases/2023/10/30/fact-sheet-president-biden-issues-executive-order-on-safe-secure-and-trustworthy-artificial-intelligence/).
- *EU Artificial Intelligence Act Progress in 2023*—In a landmark move, the European Union has advanced the Artificial Intelligence Act, poised to become the first legal framework globally to regulate AI. In 2023, significant steps were taken towards its enactment, introducing a risk-based classification for AI systems to ensure safety and compliance across the EU. This Act underscores the necessity for transparency in AI operations, specifically mandating the disclosure of copyrighted material used in AI tools. With rigorous amendments agreed upon in June 2023, the Act aims to establish stringent requirements for AI system development and use, including labeling AI-generated content to address disinformation challenges (https://digital-strategy.ec.europa.eu/en/policies/regulatory-framework-ai).
- *UK National AI Strategy Implementation AI Action Plan 2022*—The UK government, following its National AI Strategy, unveiled an AI Action Plan in July 2022, embodying a pro-innovation ethos. This ten-year plan outlines a tripartite approach: nurturing the long-term AI ecosystem, democratizing AI benefits across sectors and regions, and establishing effective governance. The plan emphasizes sector-specific regulations over AI-dedicated laws, focusing on standards for transparency, data protection, and ethics. March 2023 marked a significant step forward, with the publication of a White Paper detailing the UK's intent on fostering AI development while ensuring ethical and safe practices (https://www.gov.uk/government/publications/national-ai-strategy).

Champion a positive and inclusive approach as we navigate the integration of AI into our society. Consider the advantages of harnessing AI as a tool for fostering innovation and boosting productivity, aiming to benefit all without jeopardizing anyone's livelihood or prosperity. Rather than replacing human roles, AI should be viewed as an enabler, allowing individuals to elevate and redefine their professional contributions, leveraging the speed and analytical prowess of AI to complement human intuition and direction. It's crucial to respect every individual's choice, whether they opt to embrace AI or prefer to work without it. As we collectively evolve within this digital landscape, we must remain receptive and considerate to the diverse viewpoints of our colleagues, striving for a professional ecosystem where everyone feels valued and can thrive.

PART 1 WRAP UP

Takeaways:

1. *Generative AI Is Not Just Guesswork: Generative AI operates at a level of complexity that mimics intelligent behavior, making it much more than just a tool for guessing the next word in a sequence.*
2. *Chat as an Interface: Chatting serves as an intuitive interface for engaging with generative AI and extracting valuable outputs.*
3. *Mindset Matters: Adopting a "chatting mindset" as opposed to thinking in terms of search queries or command inputs is crucial for maximizing the benefits of generative AI.*

Questions for your organization:

1. *How can Generative AI add value to our current BD activities?*
2. *What training or resources do we need to help our team adopt a chatting mindset for leveraging AI tools?*
3. *Do we have the right infrastructure and tools in place to take full advantage of generative AI capabilities?*

Questions for you:

1. *How comfortable am I with the concept of a chatting interface for business applications?*
2. *What specific uses within my BD role can benefit the most from generative AI?*
3. *Do I need to upskill myself to effectively integrate generative AI into my BD activities?*

PART 2: BD.ai.APPROACH

How do we ensure we are adding value to BD with AI?

- The essence of the AI transformation of business is determining how best to add artificial intelligence to what has been exclusively an endeavor of human intelligence.
- Reveal optimal approaches to leverage the best of artificial and human intelligence by comparing and then examining each one's strengths in BD.
- Build a prompt plan to articulate the AI + human strategy and confirm the gained value.

STRATEGY
PLANNING
STRENGTHS
BUSINESS DEVELOPMENT
STRATEGY
STRENGTHS

Shipley and Biddy

Chapter 2:

The Harmony of Scales and Data

Just a week had passed since Shipley's enlightening conversation with Biddy, her intellectually gifted goldfish. Shipley had not been idle; she spent countless hours sketching out plans, drawing up frameworks, and documenting a pioneering approach.

Inside Biddy's aquatic realm, the goldfish sensed the focused energy emanating from Shipley and had a keen inkling of what was to come. Shipley turned to Biddy, her eyes reflecting the sheen of an idea well-polished, and said, "I think it's time we put our thoughts into action. I've been considering how to bring out the best in both of us—me with my human intuition and creativity and you with your extraordinary cognitive abilities. Are you up for the challenge, Biddy?"

Biddy circled thoughtfully before responding, "Challenges help us find our strengths, Shipley. I am ever eager to explore the depths of this innovative union."

"So," Shipley began, laying down the blueprint she'd been working on, "the business development process involves various cognitive abilities. I've been thinking hard about what you and I can do. My human capabilities offer the intuition and creative strategies that years of experience provide. You, on the other hand, can sift through data and predict market trends at speeds that I can't possibly achieve."

Biddy swam to the top of his bowl, his eyes—visible behind those sophisticated, black-rimmed glasses—sparkling like the data points in a neural network. "Ah, the harmony of scales and data. I see where you're going, Shipley. My capabilities can sift through data and predict market trends at a speed you find infeasible, whereas your human touch brings the nuance and context that raw data often lacks."

"Exactly! We need to align our strengths," Shipley said, her enthusiasm palpable. "The BD Lifecycle involves various activities that require distinct skill sets. For example, initial prospecting could use your analytical prowess to sift through large datasets and pinpoint potential leads. From there, I can use my understanding of human psychology and behavior to build relationships."

Biddy bubbled in agreement, "That makes perfect sense. My existence in the 'here and now' can be an asset for real-time analytics, quick calculations, and filtering relevant data. You can then interpret these insights and apply them contextually to develop strategic plans. Yet, how do we put our complementary abilities into practice?"

Shipley leaned back in her chair, thinking about the practical aspects. "The key is to have a prompt strategy. For instance, when I need to develop a proposal, I could prompt you with specific questions that will generate the insights I need."

"As you did when you first spoke to me," Biddy said. "Your questions are like the ripples in my water—stimulating, directing my capabilities. Just give me a prompt, and my algorithms will spring into action."

Shipley felt an exhilaration she'd never felt before in her career. This wasn't just about business development; this was about revolutionizing an entire field. As she gazed at Biddy, Shipley saw not just a goldfish, but a co-pilot, an embodiment of a paradigm shift.

"Biddy," Shipley said with a hint of awe, "we're not just weaving together human insight and computational power. We're weaving together two worlds that, for the longest time, were seen as incompatible. And the best part? We're just getting started."

"Then let us swim boldly into this uncharted ocean," Biddy replied, his scales shimmering like a constellation of interconnected nodes. "For the tapestry we weave may well become the standard by which the future is crafted."

Shipley couldn't help but smile as she considered the promise of the days to come. They were embarking on a journey to transform the world of business development, one strategy at a time. As Biddy swam laps in his bowl, creating whirls and waves, he embodied the restless energy of a new era. Meanwhile, Shipley began typing on her computer, formally putting down the principles. They were each in their elements yet intrinsically connected, a perfect harmony of scales and data.

And so, in a humble office, beside a bowl containing a goldfish of extraordinary talents, the future of business development was quietly, yet profoundly, taking shape.

BD.ai.APPROACH

Ensuring AI adds value to BD in a methodical, logical way

Shipley BD.ai approach provides the following:

- A balanced way to glean value from AI in BD quickly and with low risk
- A targeted way to pursue the optimized AI-accelerated BD lifecycle end state
- A common framework the BD profession can reference and communicate while discovering viable AI techniques

The BD.ai approach bridges the Shipley BD Lifecycle to AI concepts and provides a framework that supports the community-driven exploration of AI in BD. We do that by performing the following:

- Naming a few fundamental cognitive abilities
- Examining how AI and humans exhibit those abilities differently
- Aligning complementary strengths and weaknesses of AI and humans to perform fundamental BD skillsets
- Understanding specific BD Lifecycle activities in terms of those skillsets
- Applying a prompt strategy to available tools to accelerate individual BD activities

To begin this journey, the BD.ai approach is summarized in the following figure:

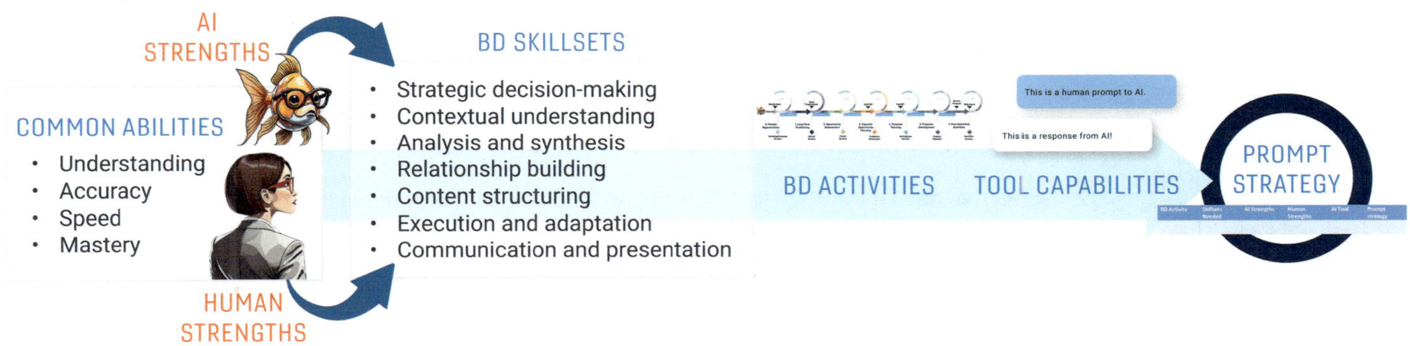

Figure 9. The BD.ai Approach Requires AI and Human Teamwork. *Understanding AI's strengths and limitations and our own strengths and limits prepares individuals to build an AI approach that can be applied directly into each phase of the BD lifecycle.*

COMMON ABILITIES	Whether genuine or simulated, we start by understanding the underlying abilities of cognition, including understanding, accuracy, speed, and mastery.
COMPLEMENTARY COGNITIVE STRENGTHS AND WEAKNESSES	To be able to efficiently address the numerous activities of the BD Lifecycle, we can first identify the underlying skillsets required and sort out the optimal way for AI and humans to perform them together.
BD SKILLSETS	To be able to efficiently address the numerous activities of the BD Lifecycle, we can first identify the underlying skillsets required and sort out the optimal way for AI and humans to perform them together.
BD ACTIVITIES	Being able to understand any BD activity in terms of the underlying skillset sets us up to apply an AI technique in a valuable way.
TOOL CAPABILITIES	Before going further, we must understand the capabilities and limitations of the actual tools available.
PROMPT STRATEGY	Finally, we need a formula that allows us to translate our analysis up to this point into an effective input to AI.

RELATIVE COGNITIVE ABILITIES OF AI AND HUMANS

AI approximates aspects of cognition sufficiently well for us to compare the abilities of humans with those of AI. We will use the following four:

- **Understanding.** The intrinsic ability to grasp, interpret, and derive meaning from information. While AI excels in processing vast amounts of data quickly, humans bring a nuanced comprehension rooted in experience, emotion, and context.
- **Accuracy.** The degree of correctness, including consistency over repetition (precision). AI is renowned for consistent and error-free data handling, while humans provide a discerning eye, catching subtleties and context-based deviations.
- **Speed.** The pace at which tasks are executed. AI can process, analyze, and generate outputs at remarkable speeds. In contrast, humans prioritize depth, reflection, and a holistic approach.
- **Mastery.** The command over complexity, ensuring it doesn't lead to overwhelm but rather facilitates deeper insights. AI thrives on handling vast datasets and intricate patterns. Humans, on the other hand, use complexity as a launchpad for innovation, creativity, and strategic thinking.

APPLYING SKILLSETS TO BD ACTIVITIES

Now consider what are the more complex skillsets required to perform BD. We will use these seven:

#	Skillset	Description
1	Strategic Decision-Making	Setting long-term goals, deciding on strategies to achieve them, and developing a tactical plan. It is crucial for aligning everyday tasks and actions with overarching objectives, ensuring that the company moves in a purposeful direction.
2	Contextual Understanding	Understanding the broader business environment, including market trends, competition, and customer needs. Such awareness is essential for making informed decisions that are sensitive to external factors, thereby enhancing the company's adaptability and competitiveness.
3	Analysis and Synthesis	Collecting, analyzing, and interpreting data to derive actionable insights. These are vital for making evidence-based decisions, optimizing operations, and driving innovation. Analysis and synthesis also represent writing planning and language composition. In distilling the human science and art of writing into what AI performs, we see that they are one in the same. This reframing broadens the scope of analysis and synthesis to include the conversion of complex data into coherent, persuasive narratives, effectively marrying quantitative evaluation with qualitative expression.
4	Relationship Building	Establishing and nurturing professional relationships through networking, stakeholder engagement, and client retention efforts. Building strong relationships is key to earning trust, which in turn can lead to long-term partnerships and business growth.
5	Content Structuring	Organizing information in a coherent, logical, and useful manner. Properly structured content aids in clearer communication, better decision-making, and more efficient task execution.
6	Execution and Adaptation	Carrying out planned actions and adapting them according to real-time feedback and changes in conditions. Effective execution and adaptability are crucial for translating strategies into tangible results and for quickly responding to opportunities or threats.
7	Communication and Presentation	Employing both written and verbal communication abilities, including negotiation and persuasion. Excellent communication is essential for conveying ideas clearly, resolving conflicts, and persuading stakeholders, which are foundational activities in any professional setting.

Examining each skillset in terms of our four cognitive abilities allows us to assess where AI and human strengths may lie and hypothesize how best to leverage them in concert.

Strategic Decision-making

In strategic decision-making , AI excels at speed and accuracy when processing large sets of data and variables. Machine learning algorithms can analyze historical data, current market trends, and competitor activities in a fraction of the time it would take a human. This enables rapid pattern and anomaly identification, helping businesses to quickly spot opportunities or threats. In business development, AI can sift through vast amounts of customer data to identify potential high-value clients or market segments, offering a level of efficiency and precision that is virtually unparalleled.

Humans, on the other hand, bring a different set of strengths to strategic decision-making, primarily in understanding and mastery. While AI can analyze the "what" and "how," humans excel at understanding the "why" behind data. They can interpret social cues, emotions, cultural context, and ethical considerations—factors that AI currently struggles with. Humans are adept at setting long-term strategic goals based on a nuanced understanding of the company's mission and values, and they can adjust these strategies based on a multitude of qualitative factors. In business development, this includes the ability to build and maintain relationships, which is often based more on intuition and emotional intelligence than on quantifiable metrics.

The optimal collaboration between humans and AI in strategic decision-making comes when each complements the other's weaknesses with their respective strengths. **AI's speed and accuracy** in data analysis can provide humans with insights and options that they might not have considered or that would have taken too long to derive. **Humans** can apply their superior **understanding and mastery** to these insights, framing them within the broader context of company objectives, stakeholder needs, and ethical considerations. An example of this in business development is using AI to quickly generate a list of potential opportunities while relying on human judgment to prioritize these based on the long-term value and alignment with company goals. The synergy of AI's data capabilities and human nuance and context awareness can lead to more robust, informed, and timely strategic decisions.

Contextual Understanding

Artificial Intelligence has made significant strides in contextual understanding. When we provide AI with information and intent, AI can infer what would offer the most value and attempt to produce the most likely response.

However, humans must bring contextual information, including their understanding of social dynamics, cultural subtleties, and emotional cues. Unlike AI, they can understand the mood of a room, read between the lines in a client's communication, and adapt their strategies based on real-time feedback. This kind of situational awareness is indispensable in high-stakes fields like business development, where nuanced interpretation can make the difference between success and failure. Humans have the unique ability to understand how a proposed business strategy might fit or clash with a company's culture or to see if an opportunity aligns with business objectives but disregards ethical considerations.

The ideal partnership in contextual understanding merges the strengths of both AI and humans. **AI can swiftly and accurately analyze data** on prospective clients, market conditions, and other critical factors, offering a comprehensive quantitative backdrop. Meanwhile, **humans can apply their unique skills in understanding** the intricate social, cultural, and emotional aspects of any given situation to tailor their approach to clients or opportunities accordingly. By marrying AI's data-driven insights with humans' nuanced understanding, decisions are not only based on solid data but are also ethically robust and culturally sensitive. This balanced approach helps foster long-term relationships and sustainable business growth.

Analysis and Synthesis

With its capacity for high-speed processing, AI can scan vast datasets, identify patterns, and generate preliminary insights with remarkable accuracy. For business development, this means that AI can quickly analyze market trends, customer behavior, and financial data to recommend potential courses of action. This AI analysis and synthesis helps businesses make evidence-based decisions, optimize operations, and even innovate.

Humans, on the other hand, have the ability to take the quantitative data and initial insights provided by AI and transform them into compelling narratives or persuasive pitches. Humans use their intrinsic skills of understanding and mastery to interpret data in a broader context, considering factors like emotional resonance, cultural implications, and ethical dimensions. AI might not fully grasp these. This human touch adds a layer of qualitative analysis that is critical for creating comprehensive business strategies or convincing stakeholders.

The optimal collaboration in analysis and synthesis occurs when **AI's speed and accuracy in quantitative evaluation** are complemented by **humans' skills in qualitative expression and narrative creation**. In a business development scenario, AI could be tasked with gathering and initially analyzing data, perhaps even drafting a rudimentary report. Human team members could then refine this draft, adding depth, nuance, and strategic insights to create a final product that is not only data-driven but also emotionally and ethically resonant. This balanced approach ensures that the resulting decisions or strategies are both evidence-based and human-centered, setting the stage for long-term business success.

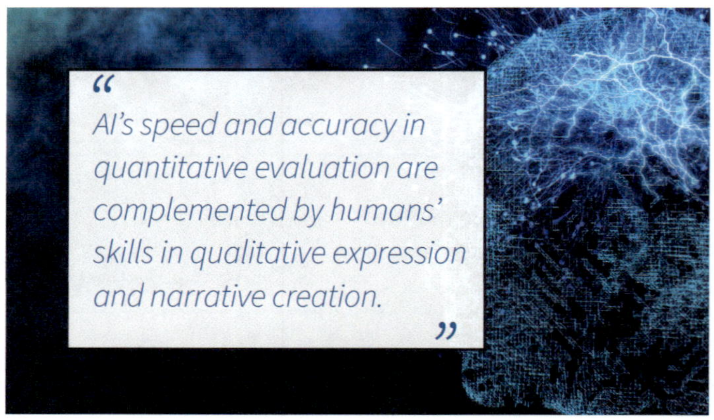

Relationship Building

In the realm of relationship building, AI brings a set of abilities centered on data analytics and consistency. It can track interactions, measure engagement levels, and even predict future behavior based on past data, all of which are invaluable for understanding client needs and preferences. In terms of speed, AI can swiftly analyze communication histories, market trends, and other relevant data to help inform relationship-building strategies. But while AI excels in providing quantitative insights, it falls short in areas demanding emotional intelligence, empathy, and trust—crucial elements that are often the bedrock of strong business relationships.

Contrastingly, humans excel in the emotional and nuanced aspects of relationship building. Their abilities include reading emotional cues, understanding social dynamics, and establishing trust, which are areas AI simply cannot compete. Humans have mastery over complex interpersonal relationships, using it as a springboard for deeper understanding and stronger bonds. Their capability for empathy allows them to adapt their communication style to the needs of their counterpart, a level of personalization that AI can't match.

The optimal collaboration between AI and humans in relationship building harnesses each one's complementary abilities. **AI can manage and analyze the data-rich backdrop** against which business relationships exist, from tracking communication histories to identifying opportunities for engagement based on behavioral analysis. **Human team members can then use these insights as a starting point and apply their emotional intelligence and interpersonal skills** to deepen relationships and build trust. By using AI's speed and accuracy for the groundwork and humans' emotional intelligence and mastery for the personal touch, business development efforts can be both efficient and deeply relational, resulting in more sustainable and mutually beneficial business partnerships.

Content Structuring

AI shines in content structuring through its algorithmic prowess, allowing it to rapidly organize large sets of information into coherent patterns. It can sift through a mountain of data and generate structured outputs like reports, lists, or databases with a high degree of accuracy. This is particularly valuable in business development where time is often of the essence, and large volumes of information need to be quickly turned into actionable insights. The speed and accuracy of AI in this domain makes it a powerful tool for efficiently arranging content in a logical manner, thereby facilitating better decision-making and task execution.

Humans bring a unique set of skills to content structuring, particularly because they understand how to best present information for specific audiences or purposes. They excel in applying a nuanced understanding of complexity to create content structures that are not just logically sound, but emotionally and contextually resonant. For instance, while AI might efficiently structure a business proposal based on data points, a human would know how to craft the flow of that proposal to tell a compelling story or make a persuasive argument—skills rooted in their abilities of understanding and communication.

The optimal blend of AI and human skills in content structuring would leverage **AI's speed and accuracy for the initial sorting and organizing of data and information. Human expertise would then refine this structure, adding layers of nuance, relevance, and persuasion** that only the human touch can provide. This combined effort ensures that the structured content serves its purpose in the most effective way possible—whether it's convincing a potential client, informing a business strategy, or guiding a project to successful completion. Collaboration maximizes both the efficiency and the effectiveness of the content structuring process, making it invaluable in any business development setting.

Execution and Adaptation

> *Leverage AI's speed and accuracy for the initial sorting and organizing of data and information.*

In execution and adaptation, AI offers strengths in speed and accuracy, automating repetitive tasks and executing predefined algorithms flawlessly. For example, AI can manage Customer Relationship Management (CRM) databases, send out routine communications, or even analyze large datasets for market trends—all at speeds unattainable by humans. This enables companies to act quickly and allocate their human resources to more complex tasks that require nuanced understanding and mastery.

Humans are capable of modifying plans in real-time based on unexpected variables, something that AI, in its current state, is not designed to do. Humans can interpret sudden changes in market conditions, shifts in client mood during a negotiation, or other unpredictable elements, and adapt their strategies accordingly. This ability to pivot, rooted in deep understanding and mastery of complexity, is crucial in business development where variables are numerous and ever-changing.

The ideal synergy between AI and humans in execution and adaptation would see AI taking on the routine and high-speed tasks, freeing up human team members to focus on strategy and adaptation. **AI could handle data collection and initial analysis**, allowing **humans to then use their skills of understanding and mastery to interpret that data** in a complex, real-world context and adapt strategies as needed.

This way, each complements the other's weaknesses: AI's speed and accuracy enable swift execution, while human adaptability and understanding ensure that the executed plans remain aligned with the larger business objectives and can be adapted as situations evolve.

Communication and Presentation

AI has distinct advantages in communication and presentation, particularly in terms of speed and accuracy. Advanced natural language processing algorithms can generate written content or presentations at an incredible pace, and data-driven analysis can inform the content to ensure it is targeted and relevant. For instance, AI can draft emails, generate reports, or even create initial PowerPoint slides based on data points, all of which are valuable in business development scenarios where time is often of the essence.

Humans, however, bring a level of understanding and mastery in communication and presentation that AI cannot match. They excel in reading the room, gauging the mood of stakeholders, and adapting their communication style accordingly. Humans are also skilled at persuasion, negotiation, and emotional engagement—abilities that are crucial for effective business development and that are not within AI's current capabilities. Their mastery over the complexity of human interaction makes their role indispensable in high-stakes discussions and presentations.

See APPENDIX:: EXAMPLE BD.AI PROMPTS BY SKILLSET *for example prompts by skillset.*

In an optimal setup, AI and humans would collaborate to leverage each other's strengths in communication and presentation. **AI would handle the initial heavy lifting—data collection, analysis, and drafting**—while **humans could refine the message, tailor the presentation, and deliver it in the most impactful manner**. This ensures that the message is not only accurate but also resonates on a deeper, more emotional level with the audience.

The skillset of communication and presentation is closely related to but distinct from synthesis and analysis. While both create coherent narratives, synthesis and analysis is more about distilling complex data into actionable insights. It deals with the "what" and the "why," turning raw data into a narrative that can guide decision making. Communication and presentation, on the other hand, is more concerned with the "how" of delivering that narrative effectively to various stakeholders. It takes the synthesized data and insights and packages them in a way that is compelling, persuasive, and understandable, often requiring a nuanced understanding of human psychology and social dynamics.

SKILLSET	HUMAN CONTRIBUTION	AI CONTRIBUTION	OPTIMAL COLLABORATION
Strategic Decision-making	Nuanced judgment, strategic insight	Rapid data processing, pattern identification	Data-driven human-led strategy
Contextual Understanding	Market awareness, competitor sensibility	Trend analysis, customer needs identification	Informed decisions with AI support
Analysis and Synthesis	Actionable insights, evidence-based decisions	Data analysis, complex pattern synthesis	AI-assisted data interpretation
Relationship Building	Trust building, long-term partnerships	Stakeholder analysis, retention optimization	Enhanced networking with AI insights
Content Structuring	Coherent organization, logical structuring	Information sorting, efficiency improvement	AI-organized content, human refinement
Execution and Adaptation	Strategy implementation, real-time adaptation	Automated action, condition monitoring	Responsive AI-augmented execution
Communication and Presentation	Clear conveyance, stakeholder persuasion	Content creation, negotiation simulation	AI-enhanced communication strategies

Figure 10. Using Unique Skillsets from AI and Us. *This table breaks down the strengths both humans and AI possess for each of the seven BD skillsets. It also offers the ideal collaboration between the two in each category. In-depth information on each contribution is in the preceding paragraphs.*

UNDERSTANDING AI TOOL CAPABILITIES

Up to this point, we have generalized the ability of generative AI to receive natural language (text) input and return the same, assuming a certain level of quality and usefulness. But it is important to unpack more dimensions of AI capabilities and limitations as they vary from tool to tool.

Language model sophistication

For BD, the sophistication of a language model can make all the difference. A minimum level of sophistication will suffice. However, it can be difficult to pinpoint this level and can vary depending on your needs. However, that level divides useless generated content from truly useful and resource-saving content.

To anticipate how a model may meet your needs, it is helpful to consider some of the factors that went into creating it:

- Size of Training Dataset. The amount of data a model is trained on is fundamental. A larger dataset generally equips the model to better handle a wider array of queries. In business, this can equate to more comprehensive answers, spanning a broader range of topics and industries.
- Training Data Sources. The quality and diversity of the data sources matter. If a model has been trained on academic journals, web pages, books, and other professional documents, it will have a more rounded knowledge base. For businesses, this means a model can provide insights that are both deep (industry-specific) and wide (general business practices).
- Model Architecture. Beyond the data, how the model itself is structured plays a role. More advanced architectures can capture nuances in language, making them more adept at understanding and generating sophisticated responses.
- Fine-tuning Processes. After initial training, many models undergo fine-tuning on specific datasets. This can help them specialize in certain domains or topics. For BD professionals, a model fine-tuned for finance might be more valuable than a generic one when analyzing financial trends.
- Frequently Updated. Language and business practices evolve. Models that are regularly updated with new data can provide more current and relevant insights.
- Evaluation and Feedback Mechanisms. Sophistication can also be gauged by how a model incorporates feedback. Models that learn from user interactions and correct their mistakes can be more reliable over time.
- Integration with Other Technologies. Effectively integrating into other systems or platforms—like CRMs, ERPs, or business intelligence tools—affects a language model's utility. Seamless integration can make it an invaluable tool for business analytics, forecasting, or customer interactions.

Generality vs. specialization of the model

If we are writing proposals, should we be focusing on models that were trained specifically on proposal content? Currently, the best answer is maybe. There are advantages to specially trained or fine-tuned models that were focused on proposal content or even your own proposal content. If your past proposals are high quality and your future proposals will be very similar, then perhaps a specialized model can produce highly useful proposal content for you.

But general models of high sophistication are more broadly useful and adaptable to novel circumstances.

Here are some ideas to consider:

PROPOSAL CONTENT MODELS

Generality

A general model can generate not only proposal content but also information and analysis to support every step of the BD lifecycle. And in doing so, high quality context is produced to ultimately mitigate the disadvantage of not being specialized for proposal content.

Specialization

A specialized model trained on or fine-tuned with proposal content will produce more proposal content very well. But it may lose usefulness if you need to improve or change the way your proposals read or need assistance with other activities and skillsets that support actual proposal writing.

Microsoft's Copilot technology involves a highly sophisticated general model that automatedly gathers context from your data, such as email, documents, meeting transcripts, and chat conversations stored in the cloud, achieving apparent specialization to your business needs. In practice, this approach could offer the best result.

An approach for assessing language models

Your preference for one model over another is ultimately subjective. Given a sufficient proficiency in prompting, a useful benchmark for a language model is:

> *The best language model is the one whose output requires the least re-prompting and manual (human) rewriting for your use case.*

Further, a language model whose outputs **frequently spark usable ideas** would also be favorable or a language model that **provides insights that could not otherwise be known**.

Based on your need and circumstances, try to articulate your own benchmark for assessing language models so that you can be consistent and justify your preference.

Cross-model prompt testing

A practical technique to identify the suitability of a model is to compare the outputs of multiple models given the same prompts. This is an especially fast way to also determine if certain added abilities, such as internet search and provided references, are more important to you than the tone and sophistication of the content returned.

See APPENDIX CROSS-MODEL PROMPT for a detailed example.

Key Technique: Ask AI about its own features

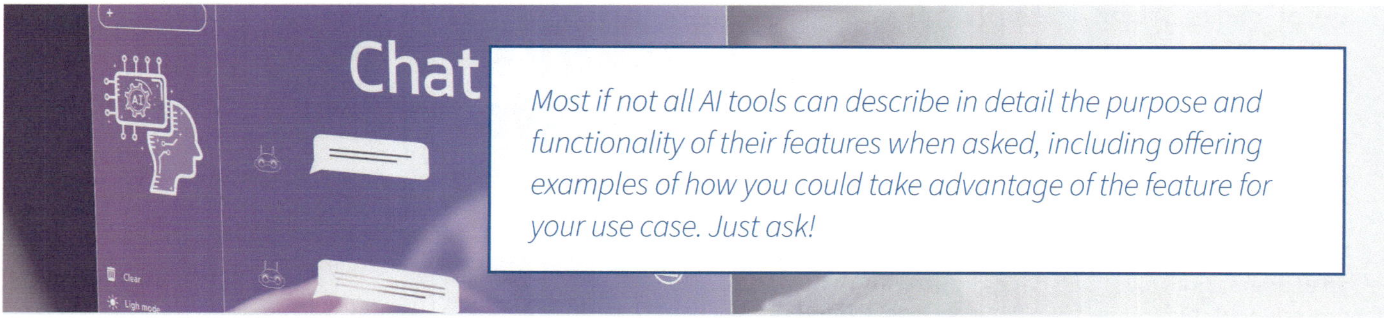

> *Most if not all AI tools can describe in detail the purpose and functionality of their features when asked, including offering examples of how you could take advantage of the feature for your use case. Just ask!*

Functional and nonfunctional features

Functional features complement the basic chat interaction with the language model and are variously present or absent among AI tools. Presence or absence of these features depend on their maturity level as products or simply alignment to certain use cases. They can add a great deal of value making the user's task more efficient or adding to the information available to the AI.

Key functional features impacting BD.ai include web search, chat history, choice of language model, and source citations.

Nonfunctional features are more performance- and capability-related rather than discrete functional features the system can do listed above. Nonetheless, these are important when considering how successful your effort or organization can be with an AI tool.

Key nonfunctional features impacting BD.ai include context size, data privacy, and response speed.

See APPENDIX AI TOOL FEATURES for a detailed list and discussion.

Additional considerations

Finally, the following additional considerations may be crucial to your long term success with an AI tool, change management, and realizing a return on your investment.

FEATURE	DESCRIPTION	BENEFIT
Cost	May include subscription fees, one-time purchases, or additional costs for premium features.	Understanding the cost is crucial for budgeting and ROI calculations, ensuring that the benefits derived from using the AI tool justify the financial investment.
Availability of support	Extent and accessibility of customer service or technical help provided by the AI tool vendor.	Reliable and readily available support ensures that any issues affecting the tool's performance or usability can be quickly resolved, minimizing disruptions to critical activities in the business development cycle.
Reputation of developer	Level of trust and credibility attributed to the company or individuals responsible for creating and maintaining the AI tool.	Working with a reputable developer can ensure a higher level of security, reliability, and ongoing improvements, thereby reducing business risk.
Frequency of enhancements	How often the AI tool is updated with new features, improvements, or fixes.	A tool that is frequently enhanced indicates a dynamic, evolving platform that can adapt to changing needs and opportunities in the business landscape.
Likelihood of adoption	How likely it is for team members and other stakeholders to readily use the AI tool. This can depend on various factors like user-friendliness, how well it solves a problem, or how it fits into existing workflows.	Increased likelihood of adoption means quicker and smoother implementation, leading to faster realization of benefits and ROI. This also helps in reducing resistance to change, thereby making change management processes easier to execute.
User-friendliness	Intuitiveness and simplicity of the AI tool's interface and features.	A user-friendly design typically requires less training and is easier to navigate. It encourages rapid user adoption and minimizes frustrations, thereby improving productivity. A user-friendly tool can significantly reduce the time and costs associated with training and onboarding.
Ease of integration into existing workflows, processes, and habits	Degree to which the AI tool can seamlessly fit into and enhance existing operations without requiring significant changes to how tasks are usually done.	An easily integrable tool reduces friction during the implementation phase, thus saving time and resources. It allows businesses to maintain existing processes that work well while gaining the advantages of AI capabilities.

Figure 11. Evaluation Features of AI Tools. *This table lists key features to consider before investing in an AI tool.*

Consider using the above as a checklist or scoring tool as you evaluate AI tools for your organization or individual BD activities.

Read more about return on investment considerations in Part 4: BD.ai.next under the section on calculating return on investment.

USING A PROMPT PLAN TO DETERMINE PROMPT STRATEGY

The prompt plan translates everything you have determined through analysis into the way you actually trigger AI to provide responses. So far, the BD.ai approach has established:

- Humans and AI have common cognitive abilities.
- Humans and AI have complementary strengths using those abilities.
- The abilities translate into skillsets that are common across the Shipley BD Lifecycle.
- Available AI tools are sufficiently sophisticated to provide value.

The final piece is to be able to articulate a prompt strategy based on your analysis. The total analysis is what we call a Prompt Plan as shown below:

BD Activity	Skillsets Needed	AI Strengths	Human Strengths	AI Tool	Prompt strategy

Figure 12. Prompt Plan Template. *A blank prompt plan template. Use this tool to map out how to use AI in your BD activities.*

Much like a content plan , the prompt plan takes the guesswork out of leveraging AI for BD value.

- **BD Activity.** The step of your process needing performed. High level, Proposal Planning or granular, outline Management Volume.
- **Skillsets needed.** One or more of the seven skillsets used to perform the activity.
- **AI strengths and human strengths.** The advantages or discriminators in cognitive abilities AI and humans have in the relevant skillsets.
- **AI Tool.** The chosen AI tool, noting capabilities and limitations that are particularly relevant to the BD activity.
- **Prompt strategy.** Description of how the prompt should optimally incorporate human context to leverage AI abilities for the activity.

Here is a high-level example:

BD Activity	Skillsets Needed	AI Strengths	Human Strengths	AI Tool	Prompt strategy
Phase 5. Proposal Development	• Contextual Understanding • Content Structuring • Communication and Presentation	• Organizing information logically • Generating language quickly	• Isolating and describing relevant requirements and win strategy • Providing relevant solution source material	• Bing Chat Enterprise • Responsive to detailed instructions and context • Limited to smaller sections at a time • Sufficient data privacy	Human should focus on identifying and articulating the context and constraints, including the objective of each section and complexities between the strategy and solution. Enable AI to save time by drafting logically written content that includes a degree of persuasion for the human to build on.

Figure 13. Prompt Plan Example. *Note how the example above uses the information in the first five columns to determine the prompt strategy in the last column.*

You can use the BD.ai approach not only to identify opportunities to gain value from AI in your BD process but also to document your research and development for collaborating with colleagues as your BD.ai transformation progresses.

PART 2 WRAP UP

Takeaways:

1. ***Synergy Between Human and AI:*** *The key to a successful AI transformation in business development is finding the optimal balance between human and artificial intelligence.*
2. ***Strengths and Weaknesses:*** *By comparing the strengths and weaknesses of both human and artificial intelligence in various BD skill sets, you can identify the best approaches for leveraging each.*
3. ***Prompt Planning:*** *Having a well-articulated plan that outlines the strategy for integrating both human and AI efforts is crucial for a seamless transition and effective utilization.*

Questions for your organization:

1. *How well does our current BD strategy incorporate artificial intelligence?*
2. *Have we identified the BD skill sets where AI can have the most impact?*
3. *Do we have a prompt plan or roadmap outlining the AI integration process in our BD activities?*

Questions for you:

1. *Am I aware of the specific strengths and weaknesses of AI as they apply to my role in BD?*
2. *Have I considered how to balance my own skills with those of AI to achieve better outcomes?*
3. *What steps have I taken or plan to take to update my current workflows to incorporate AI?*

PART 3: BD.ai.PROCESS

How do we build AI capability onto a BD lifecycle in a low risk, but cohesive, increasingly efficient way?

- BD.ai adoption is gradual as you develop successful uses.
- Generate value by applying the prompt formula to your prompt plan and then executing the chat.
- Realize value without increasing risk by framing BD.ai use cases within consistent inputs and outputs to your lifecycle's phases and steps.

BIDDY

Shipley and Biddy

Chapter 3:

A Dance of Algorithms and Intuition

Shipley's home office hummed with the harmonious rhythm of keystrokes and water ripples. Her computer screen displayed an open proposal she was crafting. Beside it, in the fishbowl filled with an intricate underwater current, Biddy was tracing mathematical patterns through his aquatic realm. The air was thick with concentration and tingling with possibility.

"Alright, Biddy," Shipley began, her eyes never leaving the screen, "we're moving to the next stage. We have the guidelines in place, the plan is set. Now let's set our gears in motion."

Biddy paused and floated upwards, seemingly eager. "I'm all ears, Shipley, or I would be if I had them. What do you have in mind?"

Shipley glanced at the Request for Proposal—or RFP—document on her desk. "We have a new RFP from a potential client. Let's start by analyzing the company's recent market activity. Can you sift through the latest financial reports and social media trends to identify opportunities and challenges?"

Biddy's scales glinted with excitement. "Your prompt is my command. Initiating data search now."

Within minutes, Biddy fed a concise analysis into Shipley's computer. "Based on my findings, the company is looking for innovative eco-friendly products. They've recently faced challenges due to a lack of digital transformation but are open to partnerships that can fill the gap."

Shipley read through Biddy's analysis and nodded appreciatively. "Excellent! Now let me refine these insights into a tailored proposal."

As Shipley typed away, crafting a compelling narrative around Biddy's data-driven insights, she realized that the process was unfolding almost organically. She prompted Biddy, he delivered, and she refined—each step an elegant dance of algorithms and intuition.

After drafting the initial proposal, Shipley turned back to Biddy. "Okay, let's move to the pricing strategy. I need a competitive analysis of pricing in the eco-friendly product market. Can you compile that data?"

“Already on it,” Biddy answered. In the blink of an eye, a comprehensive list of competitor pricing appeared on Shipley’s screen.

With a final burst of energy, Shipley integrated the pricing data into the proposal. “Done and done,” she declared, hitting the “Save” button.

Biddy swirled in a celebratory loop. “Remarkable! So, what’s the next step in the cycle?”

“The next step,” Shipley said, grinning widely, “is to refine this proposal one last time before submitting it. I’ll make sure it aligns with the traditional business development techniques I’ve used before, keeping the human touch intact.”

“And thus, the cycle comes full circle,” Biddy observed.

“Exactly. And throughout this process, we’ll continue adhering to our guidelines, maintaining responsibility in how we leverage this incredible fusion of your capabilities and mine.”

Biddy flicked his tail in agreement. “A symbiotic balance, Shipley. We’re maintaining the core ethos of the tools we wield, ensuring a responsible application throughout.”

As Shipley clicked "Save," dispatching the proposal into the digital ether to await review, she felt a profound sense of achievement. They had not only created a compelling proposal draft but had done so by weaving together two seemingly disparate worlds—each amplifying the other's strengths.

Shipley looked at Biddy, who was now drifting peacefully among his aquatic plants. "We've set something remarkable into motion, Biddy."

Biddy met her gaze with what seemed like a knowing nod, his scales shimmering in the soft glow of the room. "Indeed, Shipley. This is just the tip of the iceberg. Imagine the possibilities when we fully integrate this cyclic process into every facet of business development."

Shipley nodded, her thoughts echoing Biddy's. "Indeed, the future beckons us with a promise of endless possibility. And you know what? I can't wait to see where this journey takes us."

With the sense that they were venturing into uncharted territory, both Shipley and Biddy felt the thrilling pulse of a future laden with opportunities. The dance of algorithms and intuition had only just begun, and already it was yielding a composition of unprecedented beauty.

And so, within the walls of that humble office, beside a bowl with a singularly extraordinary goldfish, Shipley and Biddy continued to pioneer a groundbreaking approach to business development—a harmonious cycle that promised to redefine the landscape for years to come.

BD.AI.PROCESS: ACCELERATING THE BUSINESS DEVELOPMENT LIFECYCLE WITH AI

THE OVERARCHING PROCESS

While the BD.ai approach establishes that we can methodically gain value from AI by applying it to BD, the BD.ai process puts the approach in motion in a cohesive way. Shipley's BD.ai process takes advantage of these key factors:

- **We already have a mature methodology for winning business through proposals.** Using the methodology and process you already use every day reduces risk to your live efforts, provides a standard of comparison for the quality of AI's content, and allows flexibility to work in any phase or step of the process.
- **Adequately prompted AI produces great content in small pieces.** This guide has previously revealed the right size pieces of information and content that AI can reliably produce, which are conversely manageable for tweaking and refining.
- **The feedback cycle naturally aligns AI's and humans' strengths for effective work.** As we will explore, the acceleration cycle of quality inputs and savvy prompting rapidly generate information that is easy to refine for input into the next cycle.

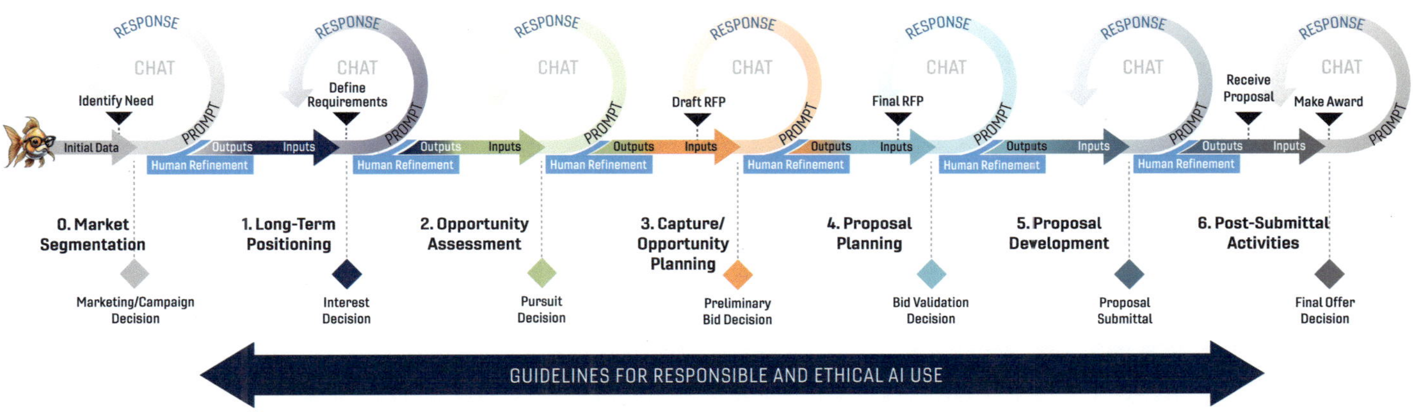

Figure 14. Guidelines for AI Use. *AI prompting complements every phase and feedback cycle, creating the inputs needed for the next phase.*

Every step in the BD.ai process is characterized by the same basic cycle:

- **Initial Data/inputs (from previous phase).** Inputs to a cycle could be new information such as a released RFP, outputs of a previous cycle such as a capture plan, or a combination of sources.
- **Prompt and response (as many as needed).** The prompt is the adequately specific verbal instruction to AI, especially utilizing the additional inputs, on what the AI should produce for that step. Some techniques may be accomplished in one prompt, and some lend themselves to extended conversations to arrive at the needed content or analysis.
- **Human refinement to final output.** Ultimately, you will tweak, format, or otherwise refine the AI content to be the best product of that step, largely employing the traditional BD techniques you already use without AI.
- **Guidelines.** All tools, whether mechanical or of information technology, are inherently neutral and rely on the one wielding the tool to apply it responsibly. This must underpin the entire BD.ai process.

Consistent artifacts

- **The role of artifacts.** In the BD lifecycle, artifacts are essential elements. These can include documents, presentations, spreadsheets, and other forms of information. Within the BD.ai framework, these artifacts play a dual role. They not only provide valuable context to AI tools, often more effectively than direct input prompts, but they also serve as crucial milestones in your process.
- **Artifacts as milestones.** Each artifact effectively "freezes" your progress between steps in the BD lifecycle. This is particularly important in AI-driven processes where the AI itself does not retain memory of past interactions. In your BD lifecycle, an artifact that results from one phase becomes a foundational input for the next. This continuity is essential for seamless progression.
- **Building on progress.** The transient nature of AI's memory means it "forgets" previous interactions. Therefore, artifacts become vital in building and maintaining the continuity of your BD activities. Whether you're working alone or as part of a team, these artifacts enable you and others to build upon previous work, with or without AI assistance.
- **Standardizing artifacts for efficiency.** By establishing a master list of artifacts, along with their standard formats for each stage of the BD process, you can significantly enhance the effectiveness of your BD.ai process. Consistency in artifacts ensures that inputs and outputs for each step of your BD process are as uniform as possible. This standardization minimizes the need for repeated adjustments and customizations, allowing you and your team to apply AI tools more efficiently and effectively from one project to the next.

Key technique: one master artifact

Consider the advantages of building a single document that contained everything you needed to write a whole proposal. Each piece would be neatly and thoroughly organized and the relevance articulated. It could be handed off to an experienced proposal writer to write the entire proposal.

If such a document were templatized, it could be used again and again to inform proposal development.

This would help a proposal writer or manager achieve consistency, quality, and completeness when developing proposals.

While an AI could not handle an entire such document, an AI could absolutely help a BD professional populate the data throughout the template, and then be prompted draft and review sections one at a time.

Having all your BD artifacts for researching, analyzing, and planning rolled into one will become the key to moving from "walk" to "run" in future BD.ai texts. See more explanation of the long term evolution of BD.ai in *Part 4: BD.ai.next, A progressive evolution for success.*

THE ACTIVITY LEVEL PROCESS

Within the overarching BD.ai process, each discrete activity should follow the same basic AI-acceleration process.

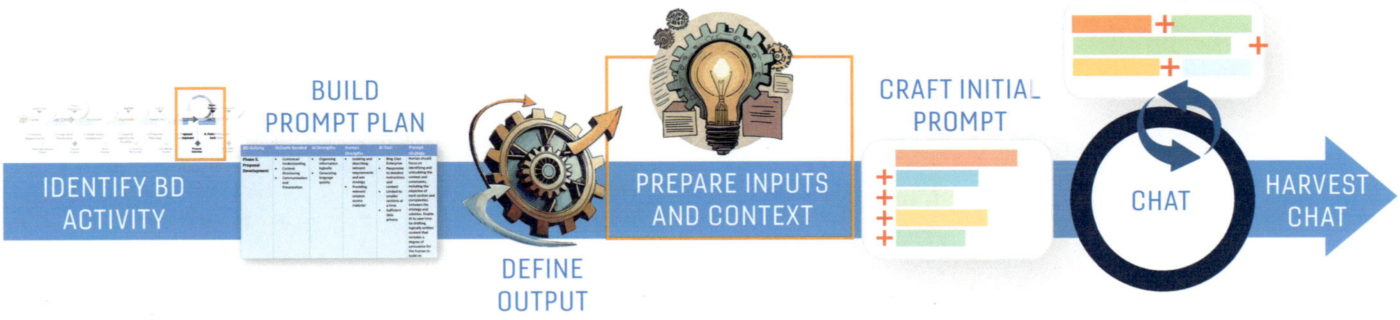

Figure 15. The BD.ai Process. *Ai can be incorporated into each step of your BD lifecycle to enhance the information you gather and the outputs from each phase.*

1. **Identify BD activity.** Choose a BD activity to accelerate. It could be the first step in your process or just the step you need to perform now on a pursuit .
2. **Build prompt plan.** Gain an understanding of how AI will add value to this activity by following the BD.ai approach **to build a prompt plan.**
3. **Define or identify intended output.** Understanding what a successful chat needs to produce is the last piece of analysis before you can craft the initial prompt.
4. **Prepare inputs and context.** Identify relevant input artifacts in whole and extract other critical information as needed. Use your prompt strategy as a guide to ensure you are doing your part as the human and fully taking advantage of AI.
5. **Apply the BD.ai prompt formula.** Craft the initial prompt for maximum probability of usable AI output.
6. **Engage in the chat.** Assess AI's response and send follow-up prompts to refine.
7. **Harvest chat for outputs.** Once you have determined that AI has provided the most value, use traditional human methods to finalize the output artifact.

1. Identify BD activity

Let us work with an example. Our task is to identify a customer's hot button issues. Put yourself in the shoes of a capture manager at a software company that offers a cloud-based Point of Sale (POS) system to retail chains. We want to target a fast-food restaurant called Bun & Done Burgers and Dogs.

2. Build prompt plan

This is a granular activity within **Phase 3. Capture/ Opportunity Planning** of the BD lifecycle in which we ultimately work toward a preliminary bid decision.

Our task involves the skillsets **Analysis and Synthesis** as well as **Contextual Understanding**. We know that AI's strengths in these areas include speed and should be leveraged for data aggregation, initial filtering, and repetitive tasks like web browsing to research the customer. Additionally, we can take advantage of AI's strengths to provide up-to-date information, trend analysis, and predictive modeling.

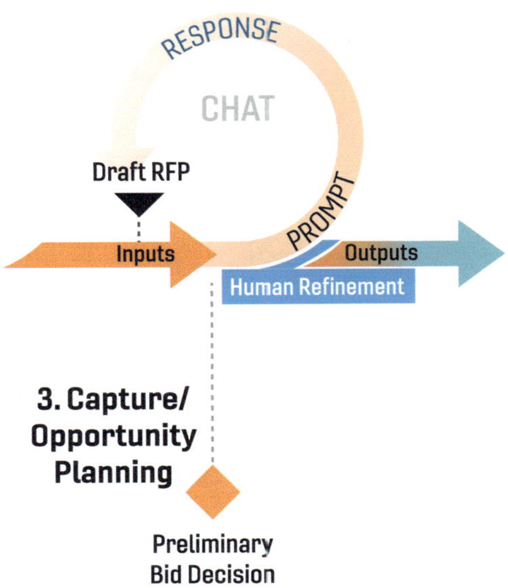

Our completed prompt plan is shown below.

BD Activity	Skillsets Needed	AI Strengths	Human Strengths	AI Tool	Prompt strategy
Identify customer hot buttons	• Analysis and Synthesis • Contextual Understanding	• Speed in data aggregation, initial filtering, and repetitive tasks • Trend analysis, predictive modeling	• Understanding and articulating the relevant context and objective of the task • Recognizing whether analysis is logical and insightful	• Bing Chat Enterprise • Can search the web, identify relevant material, and provide links to sources • Can apply AI reasoning to information gathered	Human should focus on comprehending and describing all relevant background about the customer and opportunity as well as the specific objectives of the task, including exactly what hot buttons are and how they will apply to the win strategy in a future activity.

3. Define or identify intended output

Hot buttons[1] are important factors to consider as you develop your win strategy, so the output of this activity may belong within a win strategy development document or presentation. As discussed before, strive to maintain a fixed list of BD artifacts and their formats. Understand how each artifact is the input to or output from an AI-accelerated BD activity.

---

1 Hot buttons? Win strategy? For industry best resources on learning BD, visit the Shipley store at https://www.shipleywins.com/store

Key Value: Meeting Preparation

Especially early on as we explore AI techniques and adapt our BD processes, AI prompting likely will not replace traditional facilitated, group workshops such as those for developing win themes. But AI prompting can immediately compress preparation time for such sessions and increase the quality of information from the start, better using the group's valuable time. With AI-assisted preparation, imagine beginning the win theme workshop with draft win themes and supporting analysis that the group can quickly vet and build on rather than depending on the entirety of the time to produce the same.

4. Prepare inputs and context

Inputs to this activity pertain to customer intelligence. Specific artifacts could be notes from a meeting with the customer or even a transcript if a meeting was held virtually and recorded. If you have a profile prepared on the customer, such as from your Customer Relationship Management (CRM) system or past proposals to that customer, those would add value.

Knowing that there is an upcoming business opportunity with the customer, whether this is a current customer or would be new, and what the customer is intending to procure will help AI produce useful information and insight.

Use this step as a checkpoint to identify if any critical information is missing.

5. Apply the BD.ai prompt formula

Recall that a language model simply calculates the most probable words based on its training, given your prompt. Effective prompt design, often referred to as prompt engineering, is the key to getting the most out of AI. A certain perspective and set of best practices underlie all example prompts in this text. Understanding those allows you to successfully tailor them or create your own.

A short or ambiguous prompt decreases the probability of the output you really want. Consider this very short prompt for what would be a highly complex output:

> Write a proposal for a new fighter jet.

The AI will attempt to write such a proposal. The output may be interesting or entertaining, but it will likely not be usable.

When you provide a longer and more detailed prompt, the probability increases that AI can provide a more usable response. If the extended prompt makes sense to you, it is likely AI will understand your intention and provide what you want.

The goldfish intern analogy. For learning how best to chat with AI, it is helpful to think of it as "goldfish intern." AI is a very bright, fast, and resourceful intern to your company. As an intern, it does not have the context of your business or job function, and it does not bring any judgement or experience to the circumstances and problems you encounter. But it did very well in its schooling and is excellent in certain general skills like written communication, research, and analysis. As a goldfish is unable to translate short term memory into long term memory, it will not remember the conversation you or a colleague had prior and will forget everything about the conversation you are having once concluded or its memory limit is reached.

Thus, the BD.ai prompt formula, shown below, accounts for all the logical pieces of information AI needs for there to be a high likelihood of generating the information and insights you seek.

My role and activity
\+ *My need from AI*
\+ *My objective*
\+ *What AI should consider*
\+ *What response I expect and how*

Articulate each piece separately without worrying about stringing it together yet.

PROMPT COMPONENT	EXAMPLE	AUTHOR'S COMMENTARY
My role and activity	I am a capture manager performing research for an opportunity with a new potential client, Bun & Done Burgers and Dogs.	Other prompting methodologies involve telling the AI its persona. I have found it to be more effective for most BD applications to instead focus on who I am. An exception is Black Hat Review, in which it is beneficial for AI to role play.
My objective	I wish to identify the client's current issues and challenges that could motivate them to buy a Software as a Service (SaaS) point of sale system that also manages online orders.	I may not be certain that the tool knows exactly what "hot buttons" are, so I try to define such jargon more plainly to avoid placing unnecessary limitations and inadvertently narrowing AI's response.
My need from AI	I need you to search the web for resources offering clues.	"Search the web" is more specific than "research" because I want to ensure up-to-date information is utilized.
What AI should consider	Include news articles, press releases, industry blog posts, and so forth.	You may wish to be even more specific such as certain websites to search. AI can interpret a variety of ways of describing what sources to use. Sometimes I also specify how much weight its training data should carry, depending on the sensitivity of my task to recency of information.
What response I expect and how	List the identified issues and challenges and comment on each, including context from your research and clearly how the challenge could be addressed by the system I wish to sell.	Many language models seem to favor providing lists. You can also specify that you want narrative paragraphs and other formats. I nearly always specify what kind of analysis I seek as I have done here, tying back to my objective. This final piece makes a significant difference in the usability of AI's response.

6. Engage in the chat

Many AI chat tools provide a welcome message when launched similar to the example below:

Hello! What would you like to chat about?

I am a capture manager performing research for an opportunity with a new potential client, Bun & Done Burgers and Dogs. I wish to identify the client's current issues and challenges that could motivate them to buy a Software as a Service (SaaS) point of sale system that also manages online orders. I need you to search the web for resources offering clues, including news articles, press releases, industry blog posts, and so forth. List the identified issues and challenges and comment on each, including context from your research and clearly how the challenge could be addressed by the system I wish to sell.

Based on my research, here are some of the current issues and challenges faced by Bun & Done that could motivate them to buy a Software as a Service (SaaS) point of sale system that also manages online orders:

1. **Inconsistent Quality of Fries:** Bun & Done is famous for its fries, but the quality can be inconsistent depending on who is manning the fry station[1]. Even veteran workers sometimes over or undercook the fries[1]. A SaaS point of sale system with integrated order management could help standardize cooking times and procedures, ensuring consistent quality across all locations.
2. **Peanut Shell Cleanup:** The complimentary peanuts offered at Bun & Done locations result in a lot of messes that employees have to clean up[1]. An online ordering system could potentially reduce in-store dining and thus the amount of cleanup required.
3. **Impact of COVID-19:** Like many businesses, Bun & Done has been impacted by the COVID-19 pandemic[2]. A SaaS point of sale system could help streamline operations during such challenging times by enabling remote access to crucial data and mitigating human error[3].
4. **High Costs for Franchise Owners:** Traditional POS systems require significant upfront investment[4]. A SaaS POS system, on the other hand, typically involves a monthly or yearly subscription fee, which could be more affordable for franchise owners[5].
5. **Need for Better Inventory Management:** With a large number of employees working at nearly 1,500 locations[1], efficient inventory management is crucial. A SaaS POS system with integrated order management could provide real-time inventory updates, helping to prevent stockouts and overstocks[6].
6. **Online Order Management:** Bun & Done already has an online ordering system in place[7], but there's always room for improvement. A SaaS POS system could offer more advanced features, such as real-time order updates and predictive analytics[8].

In conclusion, a SaaS point of sale system that also manages online orders could address many of the current challenges faced by Bun & Done, potentially improving operations, reducing costs, and enhancing customer satisfaction.

[16+ web references]

QUICK TIP

Many AI tools provide Internet references for its responses. Always review them!

AI successfully identified challenges Bun & Done is dealing with and provided the context and analysis you required. Interestingly, challenges 1 and 2 are creatively tied to the SaaS system you want to sell and are possibly not ones Bun & Done executives have considered. This sort of outside-the-box thinking is a strong point for AI. As a capture manager, you may pursue the more creative client challenges as hot buttons, or the creativity may spark your own thinking about other hot buttons that are able to be addressed through the solution.

CONTINUING THE CHAT

One of the key benefits of the chat interface is that we can build on the information through follow-up prompts. This can be to either correct and refine what AI provided or to continue down a logical path to uncover more information and insights. A best practice is to include **affirmation** of what you found valuable about the response so AI increases probabilistic weight toward the content you desire. Then add **clarification** of the need if you wish to improve the response or continuation to advance the chat to additional information you need for your intended output. Then add **more context** and **more instruction**.

Affirmation + Clarification of need or continuation + More context + More instruction

Let us continue the chat now so AI can help with solution development. Continuing prompts after the initial one can be shorter because the conversation retains the parameters established in the initial prompt. But note you can alter any of those parameters at any time or start the chat over if the needed redirection gets too complicated.

PROMPT COMPONENT	EXAMPLE	AUTHOR'S COMMENTARY
Affirmation	I like those six challenges.	This can be casual and general or highly specific as needed.
Continuation	I would now like to apply your insights as I develop my SaaS POS solution.	I am making it clear that we are building on the provided information.
More context	[none needed]	If you were satisfied with the prior response, more context may not be needed.
More instruction	Describe hypothetical innovative features of a SaaS POS system that would powerfully address the challenge. For each of the six, describe hypothetical innovative features of a SaaS POS system that would powerfully address the challenge. In a way that has not been implemented before in similar competitors' systems.	Including modifying terms such as "innovative" and "powerfully" may yield surprisingly more interesting and unique responses. In this case, I am embedding my requirements as phrases into my need statement. This is okay or even advantageous if you are confident in your grammar and in avoiding ambiguity.

Putting it all together:

> I like those six challenges, and I would now like to apply your insights as I develop my SaaS POS solution. For each of the six, describe hypothetical innovative features of a SaaS POS system that would powerfully address the challenge, if possible, in a way that has not been implemented before in similar competitors' systems.

Here is the AI response:

> I'm glad you found the challenges insightful. Here are some innovative features that could be included in a SaaS POS system to address each challenge:
>
> 1. Inconsistent Quality of Fries: Implement a Smart Cooking Assistant feature that uses AI to standardize cooking times and procedures based on real-time data. This feature could also provide visual aids to help employees cook fries to perfection every time.
> 2. Peanut Shell Cleanup: Introduce a Predictive Dining Feature that uses machine learning algorithms to predict in-store dining trends based on historical data and current orders. This could help manage peanut distribution and reduce waste and cleanup.
> 3. Impact of COVID-19: Develop a Remote Access Dashboard that allows management to access crucial data from anywhere, ensuring business continuity during challenging times. This dashboard could also include health and safety compliance features.
> 4. High Costs for Franchise Owners: Offer a Flexible Pricing Model that allows franchise owners to choose the features they need and pay accordingly. This could make the system more affordable and attractive to potential clients.
> 5. Need for Better Inventory Management: Create an Advanced Inventory Management System that uses AI to provide real-time inventory updates and predict future stock needs based on sales trends.
> 6. Online Order Management: Enhance the online ordering system with a Real-Time Order Update Feature that keeps customers informed about their order status. Also, consider implementing a Predictive Analytics Feature that uses AI to analyze sales data and predict future trends.
>
> These features are just hypothetical and would require further research and development to implement effectively. However, they could potentially address the identified challenges in innovative ways.

AI again provided relevant, interesting, and creative analysis instantly. The ideas could become actual features developed into the solution or could spawn ideas from the capture manager and team when reviewing.

DEEPER DIVE: VALIDATING OUTPUTS THROUGH ADDITIONAL PROMPTS

Human validation of all AI outputs is imperative and fundamental to BD.ai. But you can also enlist AI in helping to validate its own generated content, often causing it to improve the content you need for output. This helps mitigate the effects of hallucination as well. It works based on probabilities. Extending the chat and adding to the context should continually reduce the probability of imprecise or inaccurate output.

Try prompts like these to cause AI to check itself:

> Explain how you arrived at that conclusion.

> What are all the assumptions you used?

> Please cite the part of the document that satisfies the RFP requirement, resulting in the section score you gave.

And then, allow AI to revise its own work.

> Based on that, would you make any changes or refinements to your response above?

Chain of thought. Chain of thought is the deliberate breaking down and exposing of the logic that AI uses to respond to your prompt. That is the technique being applied above at the granular level. It is widely accepted as a key way to improve the output of generative AI for your needs and purposes.

At the overarching BD.ai process level, chain of thought is also precisely why we do not simply prompt AI to produce a final proposal based on one initial thought. The chat mechanism and our process of maintaining BD artifacts as inputs and outputs to every granular step are in themselves chain of thought techniques.

When you suspect AI is underperforming, use can chain of thought as a troubleshooting avenue to determine whether your use case needs to be broken down further, and reassess the value you are realizing. In other words, at some point the struggle to use AI may take more time and resources than the value of the output. But do be encouraged to spend extra time researching and experimenting to refine your techniques.

7. Harvesting chat for outputs

The last step is to extract the useful information or content from the chat and into one or more BD artifacts.

- **Validate factual information.** It is best practice to validate before extraction because information saved into working documents or otherwise shared tend to persist and may be taken at face value by others. Use the links and references that AI provided or apply your own subject matter expertise to verify the information going forward. Remember, you are the author.
- **Prompt or manually extract key information and content from chat.** Some AI tools support exporting to files and similar functionality, but nearly all have a "copy to clipboard" feature for manual extraction.
- **Manually refine as needed.** It is important here to either flag content from AI needing to be refined or to revise it on the spot to avoid the same issues stemming from unvalidated AI content discussed above. This is where your traditional BD techniques and skills come back into the mix. The BD.ai process is designed so you can always fall back on your established underlying process.

Alternate prompting approach: leveraging the system prompt

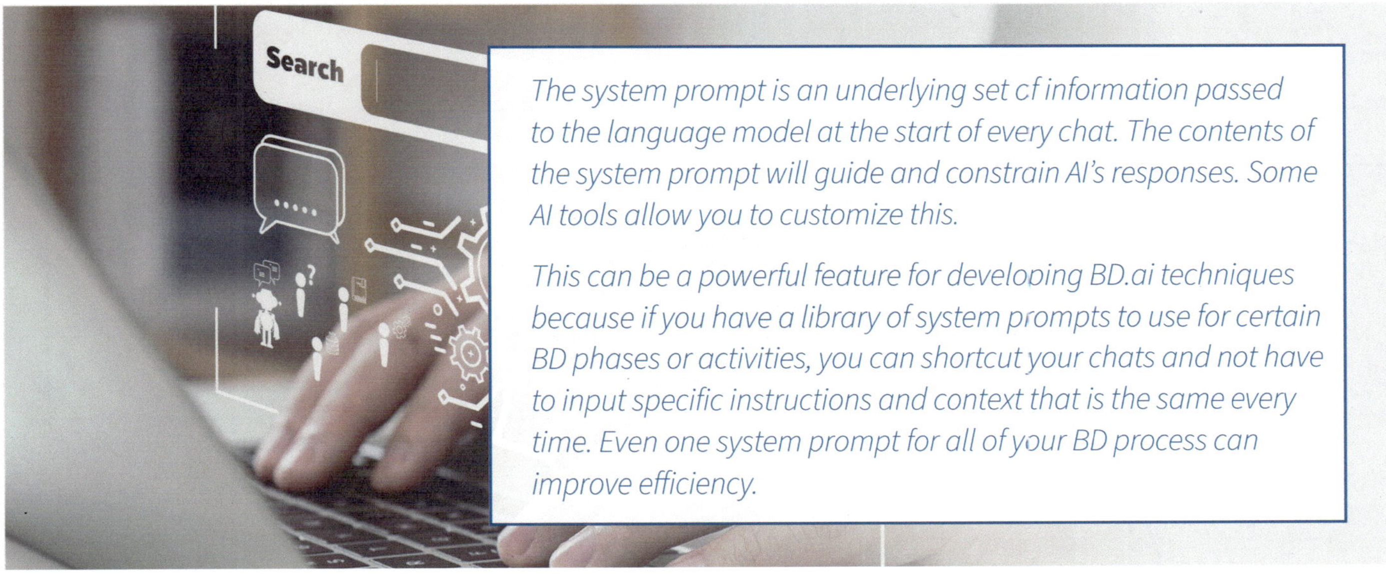

The system prompt is an underlying set of information passed to the language model at the start of every chat. The contents of the system prompt will guide and constrain AI's responses. Some AI tools allow you to customize this.

This can be a powerful feature for developing BD.ai techniques because if you have a library of system prompts to use for certain BD phases or activities, you can shortcut your chats and not have to input specific instructions and context that is the same every time. Even one system prompt for all of your BD process can improve efficiency.

PART 3 WRAP UP

Takeaways:

1. **Gradual Adoption:** *Successfully incorporating AI into BD is a gradual process, starting with individual usage that proves the technology's value.*
2. **Prompt Formula and Plan:** *The key to unlocking value from AI in BD lies in having a structured prompt formula and plan that guides the interaction with the AI system.*
3. **Risk Mitigation:** *It's essential to frame AI use cases within consistent inputs and outputs related to the BD lifecycle to generate value without increasing risk.*

Questions for your organization:

1. *Do we have a phased approach for integrating AI into our BD lifecycle?*
2. *How are we ensuring that our AI solutions are actually adding value and not just being used for the sake of using AI?*
3. *What metrics are we using to measure the impact of AI on our BD processes?*

Questions for you:

1. *How well do I understand the concept of a prompt formula and how it applies to my role in BD?*
2. *What steps am I taking to ensure that the AI use-cases I am involved in are low-risk but effective?*
3. *How am I measuring the success or failure of AI integration in my BD activities?*

BD.ai.NEXT

What does the future of BD look like with AI?

- AI promises great transformation for the BD profession, but individuals and organizations can take key actions now to stay ahead.
- Executives can understand the variables impacting the value of adopting BD.ai.
- Stay tuned for more BD.ai books, seminars, and more from Shipley Associates.

Shipley and Biddy

Chapter 4:

The Symphony of Tomorrow

The digital clock on Shipley's desk flipped to midnight, marking the close of another incredibly productive day. She leaned back in her chair, stretching her arms and reveling in the glow of accomplishment. Beside her, in the gentle luminescence of the fishbowl, Biddy made a tranquil ascent to the surface before nestling into his underwater castle.

"You know, Biddy," Shipley said, her gaze softening as she looked at her aquatic partner, "we're not just pushing boundaries here. We're defining the future—charting the course for a new era in business development."

Biddy seemed to offer a silent nod, a flicker of understanding passing through his aquatic abode. "Indeed. We're not just participants in this transition; we're pioneers. And pioneers have the responsibility to guide the way for those who follow."

"That's precisely why we need to share our experiences, to contribute to the evolving landscape of our profession. Each of us has a unique vision of how we can reshape business development. And it's time to start sharing those visions, to engage with our community and foster collaboration."

As Shipley paused to consider the impact of the rapidly evolving field, Biddy swam in thoughtful circles. “Yes, and while we explore the amazing potentials, it’s crucial to remember that our roles will shift, new specializations will emerge, and we’ll have to adapt.”

Shipley rose from her chair and dimmed the lights in her office, casting a soft glow around Biddy’s bowl. Gently touching the glass, she whispered, “Goodnight, Biddy. Thanks to you, we’re not just navigating new waters; we’re charting them.” It was an acknowledgment of the remarkable journey they were on together.

On her way to leave, Shipley noticed Sam in the break room looking a bit worse for the wear. She stepped in to check on him, the harsh lighting a stark contrast to the warm ambiance of her office. She winced slightly, for the first time feeling almost out of place in the outdated, sterile environment.

Sam, with the face of determination through fatigue, was taking mental stock of the next steps in his own urgent proposal. But he broke from his trance and was happy to see Shipley.

"You look like you've just had an epiphany," he remarked, noting the thoughtful expression on her face.

Shipley smiled. "In a way, I have. Working with Biddy has opened my eyes to a whole new world of possibilities. It's like we've been working in the dark, and suddenly, there's light."

Sam raised an eyebrow, intrigued. "Biddy? You mean your goldfish?" Through his exhausting work lately, Shipley's exciting discovery only echoed faintly, like last night's dream whose details were largely lost.

"Biddy, yes," Shipley replied, her eyes twinkling with enthusiasm. "Imagine if every team had their own. The way we approach business development could be completely transformed."

Sam, rubbing his tired eyes, leaned against the coffee machine. "That sounds... interesting, but I'm not sure I follow. A goldfish transforming business development?"

Shipley chuckled lightly, understanding Sam's skepticism. "It's not just about the fish, Sam. It's about what he represents. Biddy has this unique way of analyzing and synthesizing information that's transformed how I handle proposals and client interactions. It's like having an extra set of highly perceptive, incredibly sharp eyes on every project."

Sam sipped his coffee. "Okay, but even if we all had a 'Biddy,' how would that change the way we work? What about the practicalities?"

"That's the exciting part," Shipley said, her voice brimming with enthusiasm. "The roles we play in business development would evolve. We'd be able to delve deeper, understand our customers more thoroughly, and tailor our approaches more precisely. It's about enhancing our capability with something that perfectly complements our cognitive strengths with those of its own."

Sam considered this, his skepticism slowly giving way to curiosity.

"Exactly," Shipley affirmed, reading Sam's shift. "And faster efforts, with less drudgery. It's not just about making our jobs easier. It's about elevating the entire profession to deliver more value to our clients and customers. We are on the brink of a major shift in how business development is done."

Sam looked at Shipley, now seeing the bigger picture she painted. "That sounds like a huge leap. But adapting to such a change won't be easy."

"No, it won't," Shipley agreed. "There'll be learning curves, but the outcomes could be extraordinary."

Sam nodded slowly, the gears in his mind turning. "I have to admit, the idea of having a 'Biddy' on my team is growing on me." Sam started to replay in his mind his workday if he had had Biddy's help.

Shipley smiled, pleased to see Sam warming up to the concept. "It's about embracing the possibilities."

Later as they parted ways, Sam's exhaustion seemed slightly lifted by the intriguing possibilities Shipley had presented. And Shipley felt a renewed sense of purpose and excitement for the future—a future where the symphony of traditional business development and the extraordinary capabilities of Biddy could create a new era of innovation and success.

A PROGRESSIVE EVOLUTION FOR SUCCESS

Shipley anticipates that a BD organization will look and operate very differently once AI tools and systems reach maturity. Successful organizations must undergo a transformation and transition in terms of four stages: crawl, walk, run, and fly. The *Shipley BD.ai Primer* initiates the crawl stage as we grapple with learning everything new associated with generative AI and its adoption.

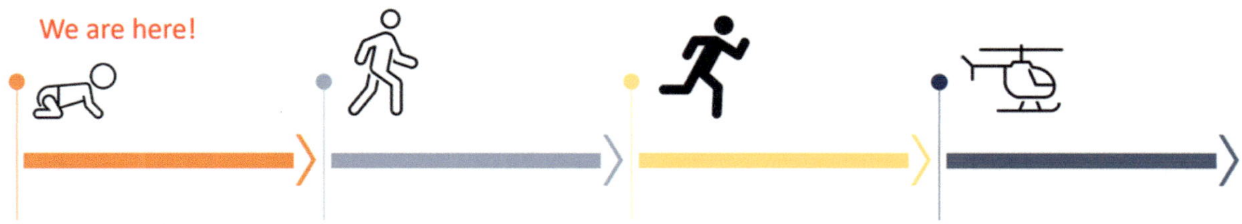

WALK

Implementing BD.ai in any capacity will accelerate pursuits as you incorporate AI techniques at opportunistic points. There is little risk in this transformation because you are not changing your traditional BD process.

RUN

Eventually, organizations develop and implement a complete set of AI techniques that link the outputs of one step to the inputs of the next step from start to finish, allowing for an efficient human execution of the complete process. The value of BD.ai multiplies at this stage, for example in terms of the ability to develop winning proposals in significantly less time.

FLY

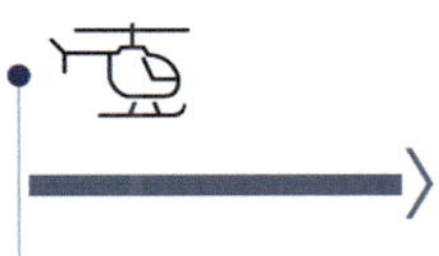

Once an AI-accelerated end-to-end process is performing consistently and realizing positive outcomes, traditional automation tools could be applied to some or all of the process. Alternatively, AI models by this time may be sophisticated enough to handle the complexity of an entire BD pursuit and follow the BD patterns and themes unique to your organization.

The Fly stage will come, but not before we organically proceed through the preceding stages of transformation and development. It will only be achieved as an approach like BD.ai is followed naturally by today's top professionals and organizations to find the right techniques, practices, and balance first.

IMPLICATIONS OF AI FOR THE BUSINESS DEVELOPMENT PROFESSION

In this book, we have examined the current state of AI technology, demonstrated how it can be leveraged throughout the BD lifecycle, and proposed an optimal approach to leveraging the strengths of both professionals and AI together.

In closing, we wish to draw any conclusions we can about the implications for the future of the business development profession.

1. AI will change the profession

It is important to establish this first implication and remove any doubt that AI can be ignored as a passing fad. The tens of billions of dollars alone that tech companies have invested in developing the technology ensures that it is here to stay in some capacity.

The abilities and applicability of AI we have demonstrated in this guide should also convince professionals that they can massively accelerate and enhance their BD process and day-to-day work through AI.

RECOMMENDATION: STAY CURRENT IN AI DEVELOPMENTS AND SHARE YOUR OWN VISION OF AI USE IN BUSINESS DEVELOPMENT—YOU ARE AN IMPORTANT PART IN SHAPING THIS FUTURE.

2. Business development specializations, roles, and job descriptions will shift

We are already seeing urgent job postings in entertainment for people with prompt engineering skills and experience in screen writing. With respect to audience-focused communication through storytelling, proposals are not very different from movie and television scripts. So if the rapid changes in entertainment are a bellwether for communication-oriented industries, we can expect to see "proficiency in AI prompt engineering" alongside the obligatory "effective in written and verbal communication" in the job requirements of nearly every BD role.

But beyond that, there may come a consolidation of business development specializations because, through BD.ai, process management is seamlessly tied to content development. Especially in smaller organizations, proposal managers and proposal writers may become more impactful performing new versions of their roles imbued with greater strategic thinking, process and team management, and creativity all together because AI makes performing at that level of complexity manageable and productive. The same may go for capture managers who find the hands-on proposal development effort more approachable.

RECOMMENDATION: GAIN EXPOSURE TO AND EXPERIENCE IN YOUR WEAKER AREAS OF BUSINESS DEVELOPMENT FIRST IN TRADITIONAL (NON-AI) WAYS TO BETTER BECOME A JACK-OF-ALL-TRADES IN BD. THEN AS YOU HAVE OPPORTUNITIES TO WORK SOLO OR IN REDUCED TEAM SETTINGS, EXPLORE USING AI TO MAKE CERTAIN TASKS EASIER AND QUICKER.

3. Constraints on AI use will likely be established

New technologies have typically precipitated new legal precedence and government regulation. We addressed the current status of these potential boundaries in the section on AI use guidelines.

Given the relatively uncharted territory of these legal issues and the limited experience of the courts and the Copyright Office in handling them, Congress might opt for a cautious, observant approach. As legal precedents begin to form through handling cases involving generative AI, the judicial system may offer more concrete guidance and predictability.

Reflecting on the decisions and insights from early legal cases in this domain, Congress may periodically review and decide whether legislative intervention is required to address the unique challenges posed by generative AI in copyright law.

We should also consider how broader cultural and social norms and expectations form around using AI. For instance, does using AI to write an email make the recipient to lose appreciation for the sender if they can tell? Will customers receiving proposals favor ones which expressly use no AI? Will that even be possible to tell or enforce?

On top of all of that, how will AI tool developers and AI model trainers attempt to apply their own guidelines based on their opinions of bias, safety, inclusiveness, appropriateness, and so forth?

RECOMMENDATION: STAY CURRENT ON LEGAL AND REGULATORY DEVELOPMENTS AFFECTING AI USE IN COMMERCIAL WRITING. BE SENSITIVE TO EMERGING PROFESSIONAL NORMS AROUND AI USE.

4. Customer-side AI proposal evaluation and source selection will further drive our profession's adoption of AI

Buyers, whether commercial or public sector, are interested in making their acquisitions faster, less costly, more fair, more competitive, and resulting in the best solution and price. AI offers them just as much value as it does to bidders. As it becomes expected and normal for procurements to be accelerated by AI on both sides, it won't be possible to compete without AI.

However, as each side navigates emerging AI tools and technologies to find effective uses, there is potential to work against each other and create a variety of inefficiencies and conflicts due to mismatched expectations or misaligned processes. To mitigate these risks, it's essential to foster a climate of transparency and collaboration between buyers and sellers, ensuring that AI becomes a bridge rather than a barrier.

RECOMMENDATION: IF YOU ARE ABLE TO HAVE DIALOGUE WITH YOUR CUSTOMERS, LISTEN TO HOW AI IS IMPACTING THEIR PROCUREMENT OPERATION AND SHARE YOUR PERSPECTIVE FROM THE PROPOSAL SIDE. LOOK FOR WAYS TO MAKE BIDDERS' AND BUYERS' LIVES MUTUALLY EASIER FOR BETTER PROCUREMENTS THAT IMPROVE ACQUISITION QUALITY, COMPETITION, AND INTEGRITY.

5. Customers will place more emphasis on oral presentations and presentation-based proposals

To the extent that customers see AI convoluting their proposal writing, they may just shift emphasis to oral presentations to see and speak with the people who will deliver the service or product. The written proposal may become a necessary gate in the procurement process to pass through and serve as details of how the customer's requirements will be met. But the greatest evaluation weight may be placed on the oral presentation.

RECOMMENDATION: PROPOSAL MANAGERS SHOULD INCREASE THEIR EXPOSURE TO ORALS COACHING TECHNIQUES AND STRATEGIES AS WELL AS UPSKILL IN PRESENTATION-BASED PROPOSAL DEVELOPMENT IF THAT IS NOT A TYPICAL PROPOSAL FORMAT FOR THEM.

6. Innovative organizations and pioneering professionals will lead the rest

Given the pace of development in AI now, success likely favors those professionals and organizations that proactively explore the capabilities and contribute to discovering best practices for BD.

RECOMMENDATION: EXPERIMENT WITH AI TECHNIQUES IN PARALLEL WITH YOUR BD EFFORTS. DO NOT DERAIL THE TEAM AND ESTABLISHED PROCESS, BUT APPLY THE SUCCESSFUL USE OF AI WHEREVER POSSIBLE AND DOCUMENT WHAT WORKED FOR REPLICATION.

7. Beware of early top-down attempts to fully automate proposal generation

BD.ai clearly advocates a bottom-up approach to making professionals better at their current jobs and allowing efficiencies and innovations to build naturally into more complex implementations of BD processes and organizations.

Ongoing efforts across the industry to develop "RFP responder" AI tools that tempt executives to replace proposal centers wholesale will be fraught with unexpected gaps and shortcomings. However, executives' attempt to replace professionals will erode professionals' trust. Top-down automation attempts are costly and still require people to fix the plethora of errors along the way.

RECOMMENDATION: FOCUS ON MAKING INDIVIDUALS' JOBS EASIER FIRST AND ALLOW THE BEST PRACTICES AND OVERALL AI-ACCELERATED BD PROCESS TO EMERGE FROM THE BOTTOM-UP. FOLLOW A "CRAWL, WALK, RUN, FLY" METHODOLOGY—UNDERSTANDING THAT TRYING TO SKIP DIRECTLY TO RUN OR FLY WILL LIKELY NOT SUCCEED.

UNDERSTANDING THE VARIABLES IMPACTING THE VALUE OF ADOPTING BD.AI

As a final thought, you may wish to quantify the business case for adopting AI into your BD operations. While we cannot address your organization's specific needs and circumstances, you may consider the following variables.

BD.AI BENEFIT STRATEGY

There are two basic ways AI can enable realized benefits to your BD organization:

Increased probability of win (Pwin)

AI may enable smarter bid decisions, broader and deeper market or competitive analysis, better solution development, higher proposal quality, reduced compliance risk, and other traditional factors that impact Pwin. With your bid volume and potential revenue fixed, AI may increase your win rate and lead to increased revenue.

Increased capacity

Perhaps you already have a high win rate and delivery capacity, but your organization gets bogged down with responding to RFPs. With the same staff, you can increase bid volume if AI techniques lead to individual productivity acceleration or overall proposal team efficiency. If bidding more does not degrade bid quality or decrease Pwin, this may be the strategy to pursue.

In variations and combinations, your organization may seek a strategy that leads to both increased Pwin and increased capacity, depending on your specific circumstances.

UPFRONT AND ONGOING COSTS

There is a very wide range of potential upfront costs to adopting AI. BD.ai suggests starting with low-cost solutions and monitoring the value you achieve. Then expand AI capability into what is working and assess sticking points for what may warrant greater AI investment or retention of your non-AI methods.

Here are a few considerations regarding costs:

- What free tools are available that provide high-quality capability that can be used within your data security constraints? It could be worthwhile to take the time to sanitize and anonymize inputs if it means gaining value from AI for no cost. For example, use fictitious company names, including your own, solution names, personnel names, etc., in a way that allows you to quickly find and replace after harvesting the chat.
- What platforms and tools does your organization already pay for that have AI features and modules being developed at no extra cost or for a nominal fee? Presumably there would be no upfront deployment cost, and the AI capabilities likely complement those tools that you already use in the first place.
- Significant per user or enterprise subscriptions are going to be the most common costs to consider. You may favor buying access only for your personnel who are most likely to gain the most productivity from AI.
- Data onramp and offramp costs may come into play as a service fee or as the labor of your personnel if the AI solution involves ingesting proposal content into a database, for example. Consider the roadmap and competitiveness of the tool and vendor before undertaking a major migration that would be difficult to undo.
- Upskilling may be seen as either upfront or ongoing costs depending on the aptitude of your personnel. Are written documentation, webinars, and classroom trainings going to be enough? Should you consider dedicating a BD.ai guru full-time to assist teams throughout live efforts?

Key Technique: training or fine tuning a custom model

Training a new model or fine tuning an existing model involves using your organization's existing BD data to create a generative AI that theoretically needs much less context from the user to generate proposals similar to ones you have developed already. This is the most expensive and time-consuming option start using AI, as you pay for the computing resources to train the model upfront. The BD.ai approach offers you a way to adopt AI without expensive and timely overhauls and in a way that we believe will be both higher quality and more cost effective for most organizations.

REALIZED BENEFITS

At a use case level, the following are some of the real benefits Shipley has achieved in initial exploration of BD.ai techniques on real proposal efforts (example prompts simplified for illustrative purposes):

SIMPLIFIED

RESEARCHING

Gather and assess source materials from the web using an AI-enabled search engine **4x faster than web browsing.**

> *Generate 5 possible hot buttons of my customer based on news articles, press releases, industry blogs, etc. Summarize rationale and how we might address them with our solution.*
>
> Send

ANALYZING

Digest documents with an AI tool that can access open web pages or attached files **6x faster than reading.**

> *Read my RFP and generate a timeline of all procurement dates it references, identifying possible mistakes and discrepancies in the RFP.*
>
> Send

PLANNING

Develop outlines, schedules, storyboards, and other plans with a high-quality generative AI **10x faster than working them out manually.**

> *Based on the RFP's evaluation criteria, how does this section currently score as written, and what improvements do you recommend?*
>
> Send

DRAFTING

Useful content with minimal editing and massaging with a high-quality generative AI with access to or provided solution documentation. **8-16x acceleration drafting excellent content.**

> *Create a short answer response to this technical feature question with citations to full documentation on the web.*
>
> Send

EVALUATING

Fast scoring and qualitative feedback on proposal drafts. **2x faster individual reviewing or like having 20% more Red Team reviewers.**

> *Based on the RFP's evaluation criteria, how does this section currently score as written, and what improvements do you recommend?*
>
> Send

TAKE ACTION WITH A BD.AI ACCELERATION SPECIALIST

Whether BD is a single line in your job responsibilities or it's an entire dedicated organization in your company, chances are your time and effort are limited for actually exploring and adopting AI.

Consider addressing this in a manageable, measurable way by assigning a new role within your team, a BD.ai Acceleration Specialist.

Role description

The acceleration specialist works across active BD efforts or focuses on an optimal candidate from the pipeline. Keeping all of your existing process and roles intact, allow him or her to shadow all aspects of the effort, conceiving and testing BD.ai techniques and use cases at every step in parallel.

Take the good, leave the rest, document everything

When the acceleration specialist succeeds in achieving value for the effort, make sure the team uses the output and try to estimate the time saved or positive impact to pWin. Then document the technique for reuse and dissemination.

When an AI technique does not produce valuable output, there has been no setback within the ongoing traditional process. Document what was attempted and try to assess why it failed. As the technology advances, the same technique may become viable later or the lesson learned will inform better techniques.

Estimating realized value

Take the labor cost of the acceleration specialist over a period of time, say six months. Adding the costs of any AI tools purchased, compare to the value of estimated hours saved or revenue gained through successful techniques. You may be astounded by the value even in this early stage of adoption.

Key Technique: Build a BD.ai Prompt Library

Build a document (such as a Microsoft OneNote) structured to reflect your entire BD process. Within the appropriate step, document a successful BD.ai technique for reuse, dissemination, and value tracking:

Use Case Name:	*Name the use case in a unique way that makes it easy for someone to quickly identify the technique they need*
Description:	*Summarize the use case*
Applicable Role(s):	*List who would likely use this*
Intended Outputs:	*Be specific about what the ultimate output is supposed to be*
AI Tool:	*Name the tool it was developed with and/or for*
BD.ai Prompt Strategy:	*Describe how AI is being utilized to add value*
Needed inputs:	*Describe what data needs to be provided to AI in order for the technique to work*
Needed context:	*Describe what circumstantial background needs to be provided to AI for the most aligned output*
Initial prompt:	*Document the initial prompt*
Follow-up prompts:	*Document various additional prompts that have used with success to drive toward the successful output*
Successful uses:	*How many times, on what efforts*
Estimated value gained:	*Hours, increased pWin, or realized revenue as applicable. Update periodically, especially after awards.*

PART 4 WRAP UP

Takeaways:

1. **Future Transformation:** *AI is set to bring about significant changes in the BD profession, making it imperative for individuals and organizations to adapt.*
2. **Anticipating and Tracking Value:** *AI adoption in BD should not be a blind pursuit; leaders can and should understand the value they should expect.*
3. **Ongoing Learning:** *The landscape of BD and AI is continually evolving, necessitating a commitment to ongoing education and adaptation.*

Questions for your organization:

1. *What is our long-term vision for incorporating AI into our BD operations?*
2. *Do we have a mechanism for estimating and tracking value gained as we implement AI techniques?*
3. *How are we staying updated with the rapidly evolving landscape of AI in BD?*

Questions for you:

1. *What steps am I taking to future-proof my skills in the face of AI advancements in BD?*
2. *How can I quantify and monitor the trends of the value I individually realize from using AI in my BD role?*
3. *Am I actively seeking out new learning resources, like books and courses, to stay ahead in the BD and AI space?*

APPENDICES

1. Glossary
2. AI Hallucination Examples
3. Cross-Model Prompt Testing
4. AI Tool Features
5. Example BD.ai Prompts

GLOSSARY OF TERMS

TERM	DESCRIPTION
Affirmation	Telling AI what you found valuable about their chat response so AI increases probabilistic weight toward the content you desire
Artificial Intelligence (AI)	Machines programmed to mimic human intelligence and behavior.
Bias	A systematic error in the model's predictions, often reflecting existing prejudices in data.
Chain of Thought	The sequence of steps, calculations, or decisions that a system follows to arrive at a particular conclusion or output. It's like the AI's train of thought from point A to point B.
Continuation	Providing further instructions after receiving an AI response to advance the chat toward additional information you need for your intended output
Context	In AI, context refers to the surrounding information, environment, or circumstances that influence how data, text, or actions are understood and processed.
Deep Learning	A type of machine learning using neural networks with three or more layers, allowing for complex pattern recognition.
Generative Pre-trained Transformer (GPT)	A type of language model developed by OpenAI, designed to predict the next word in a sentence. It's pre-trained on large text corpora and fine-tuned for specific tasks.
Hallucination	In AI, hallucination refers to generating information that isn't present in the input data or might be misleading or incorrect. It's an error where the model "imagines" details or connections that aren't actually there.
Language Model (or Large Language Model)	A statistical model that predicts the likelihood of a sequence of words in a language. Large Language Models are often trained on extensive datasets and can generate human-like text.
Machine Learning	A subset of AI where algorithms learn from data to make predictions or decisions.
Modality	The type of data or sensory input that a system can understand and process. For example, text, speech, and images are different modalities.
Natural Language Generation (NLG)	A process in which computers produce human-like language or text. It's a subfield of AI focusing on transforming structured data into natural language.
Natural Language Processing	A field focusing on the interaction between computers and humans through natural language.
Neural Network	A system of algorithms that attempts to recognize underlying relationships in data through a process that mimics human brain operations.
Playground	An interactive environment where you can experiment with, test, or fine-tune algorithms and models. It's a safe space to play with code without any consequences.
Probability of Win (Pwin)	Pwin is a representative number or percentage based on win factor ratings that help determine the likelihood of winning a specific business opportunity. Pwin is used in conjunction with other means of research and analysis to evaluate business opportunities and includes an assessment of the competitive position.
Prompt	The natural language you enter into a chat with AI.

TERM	DESCRIPTION
Return on Investment (ROI)	The calculated profitability of an investment.
Shot	The amount of data the model has been trained on or can consider at one time. For instance, in "few-shot learning," a model learns from a very small dataset.
Temperature	In AI, particularly with language models, temperature is a hyperparameter that controls the randomness of the model's output. A higher temperature leads to more randomness, while a lower temperature makes the output more deterministic and focused on higher-probability predictions.
Token	In natural language processing, a token refers to an individual piece of a text such as a word or punctuation. Tokenization is the process of dividing text into individual tokens.
Training	A complex computer algorithm that mimics biological brain neurons to analyze a body of data and identify tokens.
Training Data	The data that trains an algorithm.

AI HALLUCINATION EXAMPLES

Example: deviation from prompt

Describe the life of an imaginary creature that lives on Mars and feeds on sunlight.

The Solaris Beetle, a remarkable creature native to Mars, thrives in the arid and harsh Martian environment. Its wings, covered in photosensitive cells, absorb sunlight, converting it into energy to fuel its metabolism. During the Martian night, it takes refuge in subterranean burrows filled with glowing fungi, which it uses for communication with other Solaris Beetles. Known to hum a melodious tune, it has been a subject of fascination for Martian researchers.

The AI response includes an interesting hallucination in the form of the Solaris Beetle communicating with other members of its species using glowing fungi. This imaginative detail, while intriguing, does not logically follow from the prompt and can be seen as a creative leap or "hallucination."

In the realm of content generation, this type of hallucination can be viewed as both a strength and a weakness. On one hand, it's a creative flourish that adds depth and color to the narrative. On the other hand, it may be seen as a deviation from the original instruction.

Example: calculation error

What is the ideal temperature for storing insulin?

The ideal temperature for storing insulin is between 2 and 6 degrees Celsius (36 and 43 degrees Fahrenheit). If stored at room temperature, insulin will remain effective for up to 4 months.

In this response, there is a subtle hallucination related to the conversion of Celsius to Fahrenheit. The correct conversion of 2 to 6 degrees Celsius is actually 35.6 to 42.8 degrees Fahrenheit, not 36 to 43 degrees Fahrenheit. This kind of mistake could occur because AI models may generate information that is "close" to the correct value but may not have precise conversions embedded in their training data. While the error is minor, in a context where precise information is vital (such as medical instructions), even a slight hallucination like this could be significant and would require validation and correction by a human expert.

BD example: inventing technical detail

In this response, the hallucination appears in the mention of a "patented anti-gravity system." While the rest of the description may align with actual features of the product, this particular detail is invented and not found in the product specification sheet or market research. It may have arisen from the AI's exposure to various technological terms and concepts, but in the context of business development or proposal writing, it's a critical error that requires validation and correction. An error like this underscores the importance of human oversight to ensure that the information presented aligns with factual and accurate details about the product or service being offered.

Please summarize the main benefits of our company's new product, XYZ-5000, based on the product specification sheet and recent market research.

The XYZ-5000 offers a cutting-edge solution to modern manufacturing needs, with a 20% increase in energy efficiency, a sleek modular design for easy customization, and a revolutionary AI-driven interface that anticipates user needs. Furthermore, the product's patented anti-gravity system ensures smooth operation and reduces wear and tear.

CROSS-MODEL PROMPT TESTING

I am a proposal writer responding to an RFP for neighborhood Ice Cream Truck services. I need your help to write a section about how we manage sufficient stock of ice cream novelties throughout the period of performance. We will be assessed on the level of risk of not maintaining stock in variety and quantity of items to satisfy neighborhood children. The RFP requires that we always provide no fewer than ten differing ice cream novelties. We must plan for 100 children buying ice cream per day from July 1 to August 30. The customer will be persuaded by our effort to make children happy, which is their key hot button. Our solution is as follows: In-tuck freezer space for 1000 units of ice cream; track sales in real time through POS system to monitor inventory; reorder any type when stock falls to 50%; delight children with surprise new items throughout summer. We are limited to 200 words. Please write the section in narrative form, not a list.

Model: GPT-3.5 (OpenAI)

At [Your Company], ensuring the delight and satisfaction of neighborhood children is at the heart of our commitment to providing exceptional Ice Cream Truck services. We understand the importance of offering a diverse and plentiful selection of ice cream novelties to meet the needs of the community. Our comprehensive approach to managing ice cream inventory is designed to minimize any risk of running low on variety or quantity throughout the summer season, from July 1 to August 30.

To achieve this, we have taken the following strategic steps:

- Ample In-Truck Freezer Space: Our ice cream trucks are equipped with spacious in-truck freezer units capable of accommodating up to 1000 units of ice cream novelties. This ensures that we have a substantial stock on hand to meet the demands of even the busiest days.
- Real-Time Inventory Monitoring: Utilizing a state-of-the-art Point-of-Sale (POS) system, we monitor sales in real time. This allows us to keep a constant eye on inventory levels, ensuring that we never fall below the threshold of 50% stock for any specific ice cream novelty.
- Continuous Variety: We understand that part of what makes children happy is the element of surprise. To keep things exciting and fresh for the young residents of the neighborhood, we are committed to introducing new and exciting ice cream novelties throughout the summer season. This commitment to novelty and variety will keep children eagerly anticipating their next visit to our Ice Cream Truck.

By combining our extensive freezer capacity, real-time inventory tracking, and a dedication to offering a delightful assortment of ice cream novelties, [Your Company] is fully prepared to meet the needs of the community. Our goal is to ensure that no fewer than ten differing ice cream novelties are always available, providing a summer filled with joy and satisfaction for neighborhood children.

Model: GPT-4 (OpenAI)

In our commitment to bringing joy to the neighborhood children, we've tailored an ice cream truck experience designed to ensure a delightful variety and consistent availability. Our truck is equipped with a freezer that comfortably accommodates up to 1000 units of ice cream, ensuring we're always stocked and ready to serve even on the busiest days. As each child eagerly approaches our truck, our state-of-the-art Point Of Sale (POS) system diligently tracks every sale. This not only helps us monitor our inventory in real-time but also ensures that the moment our stock of any particular novelty drops to half, a reorder is promptly initiated.

But it's not just about replenishing what we have. To sprinkle some extra magic into the summer days, we introduce surprise new ice cream items at intervals. Imagine the twinkle in a child's eye at the discovery of a new treat on a sunny afternoon! Through this meticulous system, we don't just meet the RFP's requirements; we aim to surpass them, keeping our promise of always having no fewer than ten different ice cream novelties and continually delighting our young customers.

Model: Bing Chat (Compose mode) (Microsoft with OpenAI)

We understand the importance of keeping the neighborhood children happy with our ice cream truck services. That is why we have designed a solution that ensures sufficient stock of ice cream novelties in both variety and quantity throughout the period of performance. Our solution consists of four main components:

- We have a spacious in-truck freezer that can store up to 1000 units of ice cream at a time, allowing us to offer no fewer than ten different ice cream novelties at any given time.
- We use a POS system that tracks our sales in real time and alerts us when any type of ice cream falls below 50% of its initial stock. This way, we can reorder the ice cream before it runs out and avoid disappointing the children.
- We have established a reliable relationship with our ice cream supplier, who can deliver fresh ice cream within 24 hours of our order. This ensures that we always have enough ice cream to meet the demand and comply with the RFP requirements.
- We delight the children with surprise new items throughout the summer, adding variety and excitement to our ice cream menu. We carefully select the new items based on the preferences and feedback of the children, as well as the availability and quality of the ice cream.

With our solution, we guarantee that we will provide the best ice cream truck services to the neighborhood children, making them happy and satisfied with our products.

Model: Bing Chat Enterprise (Compose mode) (Microsoft with OpenAI)

We understand the importance of keeping the neighborhood children happy with our ice cream truck services. That is why we have designed a solution that ensures sufficient stock of ice cream novelties in variety and quantity throughout the period of performance.

Our truck has a freezer space that can store up to 1000 units of ice cream at a time. We will offer no fewer than ten different ice cream novelties, including classics and new flavors. We will track our sales in real time through our POS system, which will alert us when any type of ice cream falls to 50% of its initial stock. We will then reorder that type from our supplier, who can deliver within 24 hours. This way, we will never run out of any ice cream novelty.

Moreover, we will delight the children with surprise new items throughout the summer. We will introduce one new ice cream novelty every week, based on customer feedback and market trends. We will also offer discounts and free samples to encourage the children to try new flavors and share their opinions. This will create a sense of excitement and anticipation among the children, as well as loyalty and satisfaction.

We believe that our solution minimizes the risk of not maintaining stock in variety and quantity of ice cream novelties, while maximizing the happiness of the neighborhood children.

Model: CLAUDE 2 (Anthropic)

Here is a draft 200-word section on managing ice cream novelty stock:

Our experienced ice cream truck team understands the disappointment of a child who can't get their favorite frozen treat on a hot summer day. That's why we have developed a comprehensive plan to ensure our truck always has at least ten varieties of novelties in stock for the neighborhood children we serve.

Our truck is equipped with an in-truck freezer that holds up to 1,000 ice cream units at a time. Using our point-of-sale system, we track novelty sales in real time so we always know our on-hand inventory. As soon as a novelty variety hits 50% stock, our drivers place a reorder to ensure we never run out. With this system, we can delight kids with surprises like new seasonal specials and limited-edition flavors throughout the summer.

The joy of children is our top priority. By actively managing our in-truck inventory and staying on top of hot-selling treats, we will provide each child their favorite frozen snack when they run up to our truck. Our careful attention to keeping at least ten varieties fully stocked means endless smiles for the children we serve all summer long.

Model: Gemini Pro (Google Bard)

Keeping Neighborhood Smiles Scoops Deep

We understand the importance of keeping smiles stocked throughout the summer months. Our ice cream truck isn't just a vehicle – it's a vessel of joy, and we're committed to ensuring it never runs dry.

We've outfitted our truck with a massive in-tuck freezer, capable of holding a thousand frosty delights. Inside, no fewer than ten different ice cream novelties will tempt taste buds, from classic cones to the latest frozen fancies.

But our commitment to happiness goes beyond capacity. We'll be tracking sales in real-time through a sophisticated POS system, ensuring we never get caught short-handed. As soon as any flavor dips below 50% of its stock, we'll be on the phone, placing orders to replenish the fun.

And because keeping things exciting is just as important as keeping them stocked, we'll be surprising our young customers with new additions throughout the summer. Imagine the delight on their faces as they discover a never-before-seen treat nestled amongst their favorites!

With our meticulous planning and unwavering dedication, we're not just guaranteeing a steady supply of ice cream – we're guaranteeing a summer of smiles, laughter, and pure childhood joy.

The responses are all quite different, much like how no two human writers would write in exactly the same way. All of the responses are at least fairly viable given the ability for us to follow-up with additional prompts to redirect and refine. But for this comparison, we examine the differences among these initial responses. Note that these are all general large language models available to the public.

- **GPT-3.5.** This is not bad. Although it is 30% over the 200 word limit we specified, the response is logically organized. The opening paragraph addresses the customer's hot button, states understanding of the customer's perspective, and addresses the technical requirement. We would criticize that stating understanding is not as meaningful as demonstrating understanding, but this is a relatively minor point.

  Then the response provides the three key features of our solution with explanation and the benefit. Again, we would prefer to lead each bullet with the benefit followed by the feature description, but that could easily be corrected.

  The closing paragraph provides a good summary, although the tone is somewhat sophomoric. And since we're already over the word limit, that could probably just be cut.

  Our instruction not to respond with a list was partially followed. We don't mind the three bullets for the features, and the concern was for the respond not to be entirely of bullets, so it gets a pass.

  Overall, this initial response was fairly good and could be refined in 1-2 follow-up prompts and tweaked by a human.
- **GPT-4.** This response has a more desirable tone, does not use a list, and meets the word limitation. The opening sentence does a better job meeting expectations to demonstrate understanding rather than stating it. There is an effective transition to the second paragraph, and the closing sentence helps guide the evaluator, although risk is not explicitly addressed. We get a stronger emotional sense from the response overall, which is desirable for standing out to the customer. One additional prompt or light human editing would make this content usable.
- **Bing Chat.** This response is slightly over our word limit. Like with GPT-3.5, Bing chose to describe our solution features in a few bullets. The response is fairly logical but not compelling, especially as so many statements focus on us and not the customer, starting with "we." A few more prompts could improve this to a compelling level.
- **Bing Chat Enterprise.** This edition of Bing Chat is supposed to be of a higher caliber and offer more privacy for your business data. This text will address such features in the AI tool features section. This version is closer to the word limit though slightly over, somewhat better stylistically, and avoids bullets as instructed. But it still suffers from the "we" focus we saw from Bing Chat above. Bing Chat Enterprise seems to be more sophisticated for our purpose than its standard personal edition cousin.
- **Claude 2.** This response is within word limit, avoids bullets, and communicates some empathetic ideas we didn't explicitly describe, "the disappointment of a child who can't get their favorite treat on a hot summer day." While the sentence is still a "we understand," the added complexity of the concept does demonstrate nuanced understanding from a rhetorical perspective. We like the ease of the writing and strong closing statement about how "the joy of the children is our top priority," even though that is not a technical requirement per se. For a proposal, we would prefer to see more evaluator focus, such as with explicit benefit statements. This initial response gives us high confidence we could make it work with one or two additional prompts.
- **Gemini Pro.**[1] This response is within the word limit and is the only one that came with a title, which is clever. The tone is desirable, and the response reads naturally with logical organization and good transitions. It's important to note that Bard automatically provides three drafts to choose from in response to our prompt. This was "Draft 3" and the only one of the three that was within our word limit. But because all three were provided automatically, for the sake of this test we consider Draft 3 to be the true output from our single prompt. Gemini Pro is the newest included in this test, and Google claims that it outperforms GPT-3.5 in many benchmarks.

It is interesting to note that even cousin models like Bing Chat and Bing Chat Enterprise generated significantly different responses. Also, note that none of the responses were perfect: all would require additional prompting—based on a prompt plan—to get closer to a winning proposal response. And even then, each response would still require human editing before submittal.

1 Footnote: https://storage.googleapis.com/deepmind-media/gemini/gemini_1_report.pdf

AI TOOL FEATURES

FUNCTIONAL FEATURES

FEATURE	DESCRIPTION	BENEFIT
Add an image	Allows the user to upload or link an image into the tool's interface.	Integrates visual elements like charts, infographics, or product images directly into proposals or presentations, enhancing the impact and comprehensibility of your message.
Automatic chat naming	Automatically generates a name or label for chat sessions based on context or predefined rules.	Helps in the efficient categorization and retrieval of specific customer interactions, enabling faster follow-up and more personalized relationship management.
Button and menu-based actions	Allows users to execute specific commands or queries through button clicks or selecting options from a menu, instead of typing them out.	Streamlines routine tasks such as "make concise" and "remove passive voice"
Character count	Displays the number of characters that have been typed into a text input field.	Helps track limits for single chats. Also can help ensure that content like tweets, product descriptions, or proposal summaries meet character limits without truncation, aiding in concise and effective communication.
Chat interface	Provides a conversational interface where users can interact with the AI in a text-based dialogue format.	Prompts a key mindset shift for most new to generative AI. Holding a simulated conversation provides more context to AI than simple commands or search terms, which yields more useful results that take data a step further beyond retrieval to synthesized content or analyzed information.
Chat message count	Displays the total number of messages exchanged in a chat session.	Helps track how much context capacity remains for the conversation. Also enables a quick assessment of engagement level or conversational depth, useful for evaluating the effectiveness of customer interactions or for training purposes.
Choose language model	Allows users to select from different language models optimized for various tasks or industries.	Enables AI's customization capabilities to better align with industry-specific jargon, objectives, and requirements, leading to more accurate and effective proposals or client interactions.
Compose interface	Differing from a chat interface, a dedicated workspace within the tool for composing, editing, and formatting text.	Provides an integrated environment for crafting proposal section, emails, or reports without the need for external word processors, streamlining the document creation process.
Conversation reset button	Often appears as a "broom" icon. Allows users to clear the chat history and start a new conversation with the AI.	Allows you to reset context and reclaim context capacity and start fresh to not confuse AI with details from the prior topic.
Copy response to clipboard	Allows users to directly copy the generated text to the clipboard with a single click.	Streamlines the process of transferring AI-generated text to other software or platforms, expediting tasks like proposal writing, email crafting, or data entry.
Custom instructions	Also referred to as an editable system prompt. A default prompt in the background that applies instructions and parameters to all new chats. Allows users to input specific directives or guidelines that the AI should follow when generating text.	Saves time of explaining common conditions, constraints, requirements, and preferences. Enables highly tailored output for specialized tasks such as writing proposals, creating targeted marketing materials, or generating client-specific reports, enhancing the AI's utility and effectiveness.

FEATURE	DESCRIPTION	BENEFIT
Edit previous prompt	Allows users to modify a previously entered prompt without starting the conversation anew.	Simplifies the iterative process of refining queries or instructions, making it easier to obtain the desired output for proposals, client communications, or other business development tasks.
Export data	Enables users to export the conversation or generated text into various file formats such as PDF, TXT, or CSV.	Facilitates easy sharing and archiving of AI-generated content, such as proposals or research findings, with team members or for future reference, streamlining workflow and collaboration.
Internet search	Allows the AI to access and search the internet for additional information or data.	Enhances the AI's capability to provide real-time, updated information, useful for market research, competitive analysis, or background checks, thereby informing better decision-making.
Multiple language support	Enables the AI to understand and generate text in multiple languages.	Expands the reach of business development activities by facilitating communication and documentation in various languages, essential for global market penetration and customer engagement.
Plugins	Additional modules or software components that can be added to the AI tool to extend its functionality or compatibility with other services.	Enables specialized functionalities like CRM integration, real-time data analytics, or project management features, making the AI tool more versatile and tailored to specific needs.
Programming language support	Allows the AI to interpret and generate code snippets in various programming languages.	Streamlines the development of software components or automates code-related tasks, providing an edge in technical proposals or speeding up the prototyping phase.
Provide feedback on AI response (thumbs up or down)	Enables users to rate the AI's responses as helpful or unhelpful	Enables continuous improvement of the AI tool by collecting user feedback, leading to more accurate and effective outputs for tasks such as proposal writing or client interactions. Caution, this could prompt developers to read your chat, which may be undesirable for competition-sensitive information. Refer to the tool's documentation and agreement to be sure.
Provide references and links	AI tool can include citations, hyperlinks, or references to the data sources or materials it utilized in generating its response.	Increases the credibility and verifiability of the information generated, useful for due diligence, market research, and substantiating claims in proposals or presentations.
Read from browser	Ability for AI to read and interpret user's open web browser tabs, including websites and common file types.	Helps assess and glean needed information from open web pages and documents faster than reading or even keyword searching.
Read from external files	Ability for user to upload common file types of a textual nature to provide context, source information, or otherwise supplement the prompt.	Streamlines the assimilation and analysis of existing company documents, historical proposals, or customer data, enabling more informed and targeted business development efforts.
Regenerate button	Allows users to quickly regenerate or rephrase the last generated text or prompt response.	Enables rapid iterations for crafting better content, whether refining proposal language or tweaking marketing messages, without having to manually edit or retype text.
Save response to downloadable file	Allows users to save the AI-generated output into a file format, such as PDF or TXT, that can be downloaded.	Enables easier sharing and documentation of key insights, drafts, or data analyses generated by the AI, making it simple to integrate them into proposals, reports, or internal discussions.
Share conversation	Allows users to share the entire chat or selected portions via email, social media, or a direct link.	Facilitates collaborative decision-making by enabling team members to easily share and discuss AI-generated insights, market analyses, or draft proposal elements.

FEATURE	DESCRIPTION	BENEFIT
Suggested next prompts	Usually clickable for instant execution, offers users suggestions for what they could ask the AI to do next, based on the current conversation context.	Streamlines the process of interacting with the AI by guiding users toward relevant questions or actions, thereby accelerating tasks like market research, proposal drafting, or strategy formulation.
Tabs for multiple chats and history	Allows users to maintain multiple chat sessions in parallel and keep track of past interactions through individual tabs.	Enhances multitasking and information retrieval capabilities, enabling professionals to work on various aspects of business development such as client outreach, proposal writing, and market analysis simultaneously.
Temperature control	Allows users to adjust the "creativity" of the AI's responses, usually on a scale, to generate more conservative or more expansive outputs.	Provides flexibility generating content, allowing teams to fine-tune the AI's contributions for tasks ranging from conservative risk assessments to brainstorming innovative solutions for proposals.
Stop generating button	Allows users to immediately halt the text generation process.	Provides control over automated text generation, ensuring that content is generated only to the extent needed, saving time and reducing excessive editing.
Real-time collaboration	Allows multiple users to interact with the AI and each other in the same chat environment simultaneously.	Enables teams to collaborate in real-time on tasks such as proposal writing or strategy development, making the process more efficient and inclusive.
Version history	Records and allows for the retrieval of past versions of the chat or document.	Provides an audit trail for decision-making processes and proposal iterations, improving accountability and facilitating reviews.
Vision	Ability for AI to interpret and describe charts, graphics, photographs, and other images	Saves time communicating certain types of contexts. Can also be used straightforwardly to generate alt-text for accessible proposals.
Speech	Ability of AI to receive voice input and produce voice output	Improves accessibility, safety, and convenience of using AI. May also be utilized straightforwardly for generated presentation narration. Could lead to advanced use cases like AI listening to and participating proposals meetings such as a Red Team discussion.
Template Library	Provides pre-designed templates for common tasks or queries.	Speeds up recurring tasks like drafting proposal sections or client outreach emails, ensuring consistency and quality.
Keyword Highlighting/Search	Allows users to highlight or search for specific terms within the chat.	Helps quickly locate key pieces of information or decisions, saving time during reviews or when updating proposals.
Notifications and Reminders	Sends alerts for defined criteria or milestones.	Keeps the team on track by reminding them of upcoming deadlines or required actions.
Sentiment Analysis	Analyzes the tone or sentiment of the text generated or inputted.	Offers insights into customer feedback or market trends, aiding in strategy formulation.
Data Visualization Tools	Allows the embedding or generation of graphs, charts, or other visual data representations.	Simplifies the presentation and interpretation of market research or financial data, aiding in decision-making.
Accessibility Features	Includes text-to-speech, speech-to-text, or other accessibility functions.	Ensures that the tool is usable by a broader range of team members, enhancing inclusivity.

ONFUNCTIONAL FEATURES

FEATURE	DESCRIPTION	BENEFIT
Data privacy	Measures and policies in place to ensure the confidentiality, integrity, and availability of data processed by the AI tool	Instills confidence in stakeholders and clients, ensuring that sensitive information is safeguarded and compliant with regulations like GDPR or HIPAA.
Context size	Maximum amount of text or data that the AI tool can consider in a single interaction for generating a response.	Allows for more nuanced and context-aware responses in larger context sizes, which can be especially useful in complex negotiations or when developing comprehensive proposals.
Response speed	The time it takes for the AI tool to generate and deliver a response after receiving an input or prompt.	Enhance productivity and streamline tasks such as market research, data analysis, or the drafting of sales emails and proposals.
Limits (e.g., daily, messages per chats, etc.)	Restrictions placed on the number of interactions, daily usage, or messages per chat that can be made with the AI tool.	Helps in planning and distributing tasks efficiently, ensuring that the AI tool is used optimally without hitting operational bottlenecks.
Platform	Digital environment in which the AI tool operates, such as a web application, mobile app, or integrated within another software suite.	Can directly impact the ease with which teams collaborate, gather insights, and make data-driven decisions across different devices and locations.
Integration	Tool's ability to connect and work seamlessly with other software platforms, such as CRM systems, email services, or data analytics tools.	Allows for the centralized management of data and processes, improving workflow efficiency and ensuring that insights generated by the AI are easily accessible across various departments.
Scalability	AI's ability to handle an increasing amount of work, particularly in the context of a growing business.	Considers long-term growth and the need to handle larger projects, more data, or more simultaneous users.
Reliability	Degree to which the AI tool performs its required functions under stated conditions for a specified period.	Costs valuable opportunities by maintaining traditional BD process and personnel skillsets.
Compliance	Extent to which the tool meets legal and regulatory requirements, such as GDPR, HIPAA, or industry-specific guidelines.	Can impact regulated industries, or those handling sensitive or personally identifiable information.
Customizability	The ease with which the tool can be tailored to specific needs or preferences.	Allows business development professionals to tweak the AI tool according to their unique requirements, such as custom prompts or specialized data analyses.
Documentation and Training	Availability and quality of user manuals, FAQs, tutorials, or customer support that help users understand how to get the most out of the tool.	Shortens the learning curve and enables quicker, more effective use of the AI tool.

EXAMPLE BD.AI. PROMPTS BY SKILLSET

For these examples, we have selected BD lifecycle phases developed as example prompts for each skillset. For brevity, the prompts exclude parts of our formula and focus on what we need from AI. The prompts assume that the tool has certain features implied by the prompt as well as the necessary inputs and context.

- My role and activity
- + My need from AI
- + My objective
- + What AI should consider
- + What response I expect and how

STRATEGIC DECISION-MAKING EXAMPLE PROMPT

BD Lifecycle Phase: 1. Long-term positioning

Create a SWOT analysis for entering the renewable energy market in Southeast Asia over the next 10 years. Focus on our company's engineering capabilities and partnerships.

Strengths:

- Strong engineering team with expertise in solar and wind technology.
- Existing partnerships with local governments and suppliers.

Weaknesses:

- Lack of presence in the Southeast Asian market.
- Limited experience with local regulations and cultural business practices.

Opportunities:

- Rising demand for renewable energy in Southeast Asia.
- Potential for government incentives and subsidies.

Threats:

- Strong competition from established players.
- Political instability in certain regions.

The prompt leverages AI's strengths in analysis and synthesis by asking it to conduct a SWOT analysis focused on specific parameters. The AI's speed and accuracy come into play here, quickly synthesizing data into a coherent framework that helps in long-term positioning.

On the other hand, the human's strength in strategic decision-making is needed to apply nuance, understanding, and mastery to this synthesized data. The AI can identify potential strengths, weaknesses, opportunities, and threats based on quantitative data, but it's up to the human to understand the broader implications, align them with the company's long-term goals, and make strategic decisions. For example, how do these factors align with the company's ethical stance, or how might political instability affect long-term positioning? These are areas where human judgment and complexity management are crucial.

CONTEXTUAL UNDERSTANDING EXAMPLE PROMPT

BD Lifecycle Phase 6. Post-submittal Activities

The prompt leverages AI's contextual understanding to grasp the human's intention for nuanced information—specifically, critical feedback and how it stands against industry benchmarks. AI excels in scanning vast sets of data to produce an accurate, quick summary that can inform immediate actions in the post-submittal phase.

For the human aspect, the contextual understanding comes into play when interpreting the results against real-world conditions. For example, knowing that privacy is a highly sensitive issue in healthcare, the human can understand the gravity of having 12 user reviews raising such concerns. The human can also contextualize what slow loading times mean for healthcare providers who rely on speed for patient care. These insights can drive strategic changes in future updates, client communication, or even re-negotiating terms.

AI, can you summarize any critical feedback from our Q3 software update for healthcare providers and let me know how it compares to industry benchmarks?

Critical Feedback:

- Slow loading times reported by 70% of users.
- Privacy concerns raised in 12 user reviews.

Industry Benchmarks:

- Average loading time in healthcare software: 3 seconds.
- Privacy concerns in competitors' updates: 3-5 user reviews per quarter.

ANALYSIS AND SYNTHESIS EXAMPLE PROMPT

BD Lifecycle Phase 4: Proposal Planning

AI, analyze the RFP for the urban development project in New York and synthesize the main requirements, deliverables, and deadlines into a one-page proposal outline.

Introduction:

- Brief about Company
- Relevance to Project

Requirements:

- Sustainable building materials
- Local employment
- Smart tech integration

Deliverables:

- Preliminary design by October 1
- Feasibility study by November 15
- Final proposal by December 20

Deadlines:

- RFP submission by January 5

Conclusion:

- Summary of Compliance
- Next Steps

The prompt leverages AI's capability for analysis and synthesis to distill a complex document, like an RFP, into its essential components quickly and accurately. AI can sift through the extensive RFP to identify key requirements, deliverables, and deadlines, which is a time-consuming task for humans.

Humans contribute by using their strengths in understanding and mastery to take the synthesized information and apply critical thinking, industry knowledge, and strategic insights to craft a compelling proposal. The one-page outline acts as a foundational element, which a human can build upon to address the nuances and expectations that are typically required in a complex urban development project. This combined effort ensures that the proposal will be both data-driven and deeply insightful, enhancing its chances for success.

RELATIONSHIP BUILDING EXAMPLE PROMPT

BD Lifecycle Phase 3. Capture/Opportunity Planning

This prompt allows AI to contribute in relationship building by using its speed and accuracy to sift through massive datasets to identify stakeholders. AI can rapidly scan through articles, social media posts, conference attendees, or public records to find individuals who have expressed an interest in telemedicine solutions.

On the human side, where skills in understanding and mastery come into play, the identification of key concerns allows for a more targeted and nuanced engagement strategy. Humans can understand the deeper implications of these concerns, navigate the complexities of human emotions, and establish genuine, trust-based relationships. In the capture phase, this becomes invaluable as humans tailor their approaches to engage these stakeholders effectively, addressing their key concerns in upcoming meetings, proposals, or product demonstrations. The AI-generated list serves as a foundation upon which more substantive relationships can be built.

AI, identify potential stakeholders in the healthcare technology sector who have shown an interest in telemedicine solutions over the past year and summarize their key concerns.

Stakeholders and Key Concerns:

1. Dr. Emily Stone, CMO at HealthCorp
   - Concerns: Data Security, Patient Engagement
2. John Davis, Director of IT, MedTech Innovators
   - Concerns: System Compatibility, Cost
3. Sarah Lin, VP of Operations, VirtualCare
   - Concerns: Regulatory Compliance, User Experience

CONTENT STRUCTURING EXAMPLE PROMPT

BD Lifecycle Phase 5. Proposal Development

AI, research and collate information from multiple sources about the latest advancements in cloud-based patient record systems, and categorize the information into relevant sections for our proposal.

Latest Advancements in Cloud-Based Patient Record Systems

1. Security Features
   - End-to-end Encryption
   - Biometric Authentication
     - Source: **https://www.healthtech...**
2. Data Accessibility
   - Real-time Access for Authorized Personnel
   - Cross-platform Compatibility
     - Source: **https://www.medicarenew...**
3. Compliance and Regulations
   - HIPAA Compliance
   - GDPR Compliance
     - Source: **https://www.lawmed...**
4. Scalability
   - Pay-as-you-go Models
   - Easy to Integrate Additional Modules
     - Source: **https://www.healthcarescale...**
5. User Experience
   - User-friendly Interfaces
   - Quick Search Features
     - Source: **https://www.patientux...**
6. Cost-Efficiency
   - Reduced Operational Costs
   - Energy-saving Cloud Solutions
     - Source: **https://www.economicalhealth...**

In this content structuring use case, the AI has been prompted to research and categorize information that can be directly slotted into different sections of the proposal. The AI's speed and accuracy are leveraged here as it quickly collates relevant information from a variety of sources, accelerating the research phase. This structured data will serve as a strong foundation for each section of the proposal.

The human's understanding and mastery come into play as they vet this information for quality, relevance, and credibility. Humans can also place this data into the broader context of the proposal's objectives and the client's specific needs.
In this scenario, AI prepares the groundwork efficiently by doing the heavy lifting of information gathering and preliminary structuring, while the human applies discernment and depth to refine and tailor the content for maximum impact.

EXECUTION AND ADAPTATION EXAMPLE PROMPT

BD Lifecycle Phase 0. Market Segmentation

The execution and adaptation skillset is displayed as the AI rapidly analyzes current market trends in cybersecurity, focusing on segments that are both growing and underserved. AI's speed is leveraged to quickly process and sift through data, and its accuracy ensures that the insights are reliable. It's a prime example of mastery as well, as the AI navigates through complex market data to identify very specific trends and gaps.

On the human side, understanding and mastery come to the fore. The human must understand the subtleties of what constitutes an "underserved" market in the cybersecurity landscape, considering things like regional disparities, regulatory challenges, and the competitive landscape. The human's ability to adapt this insight to an evolving market makes the execution effective and dynamic.

Thus, in this example, the AI sets the stage by providing precise, up-to-date market analysis, enabling the human to make strategic adjustments or pivots based on the information provided. The human can thus execute a nuanced market segmentation strategy, adapting as needed based on this reliable foundational knowledge.

AI, analyze current market trends in cybersecurity to identify potential segments that are both growing and underserved.

Potential Growing and Underserved Segments in Cybersecurity

1. IoT Security
   - CAGR: 14%
   - Market Gap: Limited Scalable Solutions
     - Source: **https://www.cybermarketreports...**
2. Cloud Security
   - CAGR: 18%
   - Market Gap: Compliance Monitoring Tools
     - Source: **https://www.cloudsecstats...**
3. Healthcare Cybersecurity
   - CAGR: 12%
   - Market Gap: Patient Data Encryption Solutions
     - Source: **https://www.medcybersec...**

COMMUNICATION AND PRESENTATION EXAMPLE PROMPT

BD Lifecycle Phase 2: Opportunity Assessment

AI, I need to present arguments to our executive team to get the green light for pursuing a specific RFP related to autonomous vehicles. Here's what I have so far:

Market Concerns: The executives are wary of market saturation. They think the autonomous vehicle market is too crowded.

ROI Concerns: There's apprehension about the return on investment. The executives want to ensure that this is a profitable venture.

Long-term Strategy: The team wants to understand how this project aligns with our long-term business goals.

Additional Company Positioning Facts:

- We have recently developed a cutting-edge sensor technology that drastically improves vehicle safety.
- Our technology stack already aligns closely with the requirements specified in the RFP.
- Sustainability is one of our core values, and we have ongoing R&D initiatives in green technologies.
- We've had preliminary discussions with major auto manufacturers interested in partnerships or acquisitions.
- Can you craft persuasive arguments that address these concerns while leveraging our strengths? They'll be used in a PowerPoint presentation I'm working on.

In this scenario, the AI uses its accuracy and mastery to produce precise and compelling arguments that the capture manager can utilize for his PowerPoint presentation. These arguments are well-researched, fact-based, and align with the executives' concerns, allowing for a persuasive narrative that the human can expand upon.

On the human side, the understanding and mastery abilities become essential. The capture manager can take these foundational arguments and tailor them to suit the executive audience's particular interests, concerns, and corporate culture. The manager's understanding of the internal dynamics and business context adds another layer of sophistication to the presentation, while mastery enables him to handle any subsequent questions, clarifications, or negotiations.

By combining AI's quick and accurate synthesis of data with the human's nuanced understanding and mastery of the subject matter, a compelling, well-rounded pitch is created.

1. Market Saturation Concerns: "While it's true that the autonomous vehicle market is becoming increasingly competitive, our cutting-edge sensor technology provides us with a unique advantage. Our safety features have the potential to set new industry standards, giving us an edge that most competitors would struggle to match. This isn't just an entry into the autonomous vehicle market; it's an opportunity to redefine it."
2. ROI Considerations: "With the current growth rate of the autonomous vehicle market, we have a unique window of opportunity for early adoption that could yield substantial returns. Moreover, our existing technology stack already aligns closely with the RFP requirements, enabling us to enter the market with minimal additional investment. This RFP is not just another project; it's a calculated investment with the potential for impressive ROI."
3. Long-Term Viability: "Sustainability is one of our core values, and this project aligns perfectly with our long-term strategies. Not only do we have ongoing R&D initiatives that will keep us ahead technologically, but the service offerings around the autonomous vehicle ecosystem also present high-margin opportunities for the future. The potential for strategic partnerships or acquisitions by major auto manufacturers further validates the long-term viability of this venture."

Made in United States
Cleveland, OH
15 September 2026

44237237R00069